Lecture Notes in Chemistry 61

Edited by:

Prof. Dr. Gaston Berthier
Université de Paris

Prof. Dr. Michael J. S. Dewar
The University of Texas

Prof. Dr. Hanns Fischer
Universität Zürich

Prof. Dr. Kenichi Fukui
Kyoto University

Prof. Dr. George G. Hall
University of Nottingham

Prof. Dr. Jürgen Hinze
Universität Bielefeld

Prof. Dr. Joshua Jortner
Tel-Aviv University

Prof. Dr. Werner Kutzelnigg
Universität Bochum

Prof. Dr. Klaus Ruedenberg
Iowa State University

Prof Dr. Jacopo Tomasi
Università di Pisa

D. Searles E. von Nagy-Felsobuki

Ab Initio Variational Calculations of Molecular Vibrational-Rotational Spectra

Springer-Verlag Berlin Heidelberg GmbH

Authors

Debra J. Searles
Research School of Chemistry
Australian National University
Canberra ACT 2600, Australia

Ellak I. von Nagy-Felsobuki*
Department of Chemistry
The University of Newcastle
Callaghan NSW 2308, Australia

*To whom correspondence should be addressed

ISBN 978-3-540-57465-1   ISBN 978-3-662-05561-8 (eBook)
DOI 10.1007/978-3-662-05561-8

© Springer-Verlag Berlin Heidelberg 1993
Originally published by Springer-Verlag Berlin Heidelberg New York in 1993

Typesetting: Camera ready by author
51/3140 - 5 4 3 2 1 0 - Printed on acid-free paper

Preface

This work had its beginnings in the early 1980s at the University of Wollongong, with significant contributions from Dr. Margret Hamilton, Professors Peter G. Burton and Greg Doherty. The emphasis was to develop computer code to solve the nuclear Schrödinger problem. For bent triatomic molecules the project was finally realized at the University of Newcastle a decade or so later, with the contribution from Ms. Feng Wang. Aspects of this work are now taught in the quantum mechanics and electron spectroscopy courses at The University of Newcastle.

Even now "complete" ab initio solutions of the time-independent Schrödinger equation is not commonplace for molecules containing four atoms or more. In fact, when using the Eckart-Watson nuclear Hamiltonian a further restriction needs to be imposed; that is, the molecule is restricted to undergoing small amplitudes of vibration. This Hamiltonian is useful for molecules containing massive nuclei and moreover, has been extremely useful in interpreting the rovibrational spectra of small molecules. Nevertheless, a number of nuclear Hamiltonians that do not embed an equilibrium geometry have become well established and are extremely successful in interpreting rovibrational spectra of floppy molecules. Furthermore, solution algorithms vary greatly from research group to research group and it is still unclear which aspects will survive the next decade. For example, even for a triatomic molecule a general form of a potential function has not yet been uncovered that will generally interpolate with accuracy and precision ab initio discrete surfaces.

This monograph hopes to encapsulate one approach (rather than "the" approach) in solving the nuclear Schrödinger equation for triatomic molecules. In doing so, it is limited. Nevertheless, it is hoped that a number of difficulties not yet resolved will become evident in the process of detailing our adopted methodology.

A number of people and bodies have given their support and/or made facilities available in order that this work could come to fruition. In particular, support from Marie-Therese Wisniowski, Paul Bernhardt, Feng Wang, John Ogilvie, Peter Burton, Greg Doherty and Margret Hamilton is greatly appreciated. We especially wish to thank Australian Academy of Science - National Science Council (ROC) for a visiting Professorial Fellowship for EvN-F; the Outside Study Programme of The University of Newcastle for support of EvN-F while on leave; Australian Research Council; Research Management Committee of The University of Newcastle; ANU Supercomputer Centre and the Computing Centre of The University of Newcastle; National Tsing Hua University and Institute for Atomic and Molecular Science (Taiwan).

Debra J. Searles and Ellak I. von Nagy-Felsobuki

TABLE OF CONTENTS

CHAPTER I

HISTORICAL REVIEW

§.1. Historical and General Considerations.

Spectroscopy can be defined as the quantitative study of radiation by dispersing its component frequencies and attributing a relative intensity to each. It is generally applied to the investigation of the behaviour of light when interacting with matter. The technique gives precise information on the atomic and molecular source of incoming and/or outgoing radiation, thereby probing the structure of matter (which make its presence known by resonant or non-processes). Since the mid 19th century it has become indispensable in structure determination and in chemical analysis.

The history of luminescence from the primitive cultures until 1900 has been documented by Harvey [1] and the historical development of quantum theory has been detailed Mehra and Rechenberg [2]. The monographs of Herzberg [3-4] still provide a useful compendium of techniques, theoretical concepts and spectral data essential to understand spectroscopy. Historical vignettes of the people and ideas, which have developed the field of rovibrational spectroscopy shall be given below, in order to place in historical context the more recent theoretical approaches dealing with ab initio variational calculations of molecular rovibrational spectra - the topic of this monograph.

Prior to the 19th the century, the study of light interacting with material occupied an important place in superstition among all cultures. Moreover, the contrast of light and darkness provided the vehicle for the Biblical account of creation; that is, first there was chaos followed by darkness and then light, the appearance of which heralded the beginning of the world.

Although the word "light" has often been used figuratively, as in Scandinavian mythology, the folklore of every people is full of legends that rationalize the emitted and reflected light observed from say, heavenly bodies. For example, the Australian Aboriginal account of the appearance of the morning and evening star is indicative of the general "oneness" and harmony that an indigenous people have with their environment. That is, the environment is an extension of themselves. The Aboriginal myths are not written records, but rather an oral history which is retold from one generation to another. The following version of the creation of the morning and evening star was told to Roland Robinson by Nalul (of the Djauan tribe) [5] and gives insight into the views of a culture that spans a 40,000 year period.

"In dreamtime a blackfellow went down to a river to have a swim and to gather some mussels. As he walked into the river he saw a bright stone lying on the bottom. He picked it up out of the water and held it, wet and shining, on the open palm of his hand, looking at it. He had only just started to look at it, when suddenly, it went away out of his hand and up into the sky.

This stone was a blackfellow named Munjarra. And Munjarra went on up to the sky and began to walk up there..... He walked over the sky this way and that, but could see no way to get down. Then Munjarra realized that he would have to stay up there. 'Oh', he said,'it must be that I have to be like Morwey the sun and burn up the earth.' But Morwey the sun heard him say this, and in a loud voice Morwey shouted,'No! I am the one who has to light up the day-time. You had better go and light up the night-time.' And Munjarra went away from Morwey the sun and waited until Morwey had gone down out of the sky. And then Munjarra came out and began to light up the night-time..... He was the first (*and last star*), burning brightly above the place where Morwey had gone out of the sky."

The Australian Aborigines associated light with heat and were able to distinguish between reflected and emitted light. Generally, the source of heat and light, such as the Sun, occupied an important place in the heavenly hierarchy. The ancient Greek sun god was Helios, who the Romans adopted in the form of Phoebus Appollo. They recognized that faint cold light (luminescence) was possible and associated it with moonlight (whose goddess was Selene or Luna).

The production of light without heat (luminescence) was generally associated with supernatural significance. It is therefore surprising that whilst the firefly is recorded in sacred books of India and China, neither the firefly nor the glowworm is mentioned in the Bible, the Talmud or the Koran. The systematic study of luminescence began with Aristole, who listed in Historia Anuimalium a number of producers of cold light.

The Dark and Middle ages did not add any new discovery or list any new phenomena associated with the production of light or its interaction with matter. Rather the few mysterious and miraculous stories which did appear, were later refuted, and so by the end of the Middle ages the understanding of light phenomena in the Western World remained in the same state as during the classic era of the Greeks and the Romans [1, 6].

During the fifteenth and sixteenth century little progress was made in rectifying the mistakes or interpretations of the ancients. In fact, the knowledge of the ancients was regurgitated, although a number of new luminous animals were discovered in the New World and so extension to the previous knowledge was collected in the works of the great naturalists. At the end of the sixteenth century Bacon, Verulam and Galilei all adopted the scientific method. Of the three, Bacon paid most attention to luminescence and was able to give the first adequate enumeration of low temperature light emission [1, 6].

Like some much of science, Newton laid the foundation for spectroscopic analysis with prism experiments. However, Wollaston first reported a characteristic spectrum from light restricted to a narrow slit. In 1814 Fraunhofer, while testing achromatic lenses, found that the continuous spectrum of the Sun yields hundreds of dark lines at specific wavelengths. He thought that the lines were characteristic of sunlight and not of the instrument used to observe them. He also noted that the dark lines in the yellow region of the spectrum coincided with the yellow line emitted by the flame of a candle. It was suggested that perhaps this experiment might be useful for chemical analysis. It was not until 1859 that such an analysis was feasible. In that year, Kirchhoff and Bunsen developed the first recording spectroscope.

Stokes, Thomson, Stewart, Kirchhoff and others suggested that certain resonant frequencies characterise each element of the periodic table. Optical spectroscopy thereafter played a pivotal role in astronomy, in chemical analysis, and, much later provided empirical evidence for testing theories of atomic and molecular structure [2].

The experimental and theoretical development of electrodynamics rested on the shoulders of Faraday and Maxwell. The actual proof that light was an electromagnetic phenomenon was provided by Hertz. It was after his demonstration in 1888 that optical theory became an integral part of electrodynamics.

Balmer, Rydberg, Ritz developed empirical formulae to deduce the spectral lines for a number of elements. Paschen used the combination principle in a systematic search for the infrared spectra of the elements. Paschen discovered the infrared series of hydrogen and Lyman the ultraviolet series. An integral part in these observations was the use of the concave diffraction grating of Rowland, which not only enlarged the resolution of wavelengths, but also avoided self-absorption by the material of the prism.

In 1900 Planck had to assume fixed packets of energy (with magnitude hv) in order to understand the energy exchange between oscillators and the radiation field (black-body radiation). As early as 1903 Thomson suggested that the energy of light is distributed as "specks" across an expanding wavefront. However, it was until 1906 that the general scientific community, via the work of Einstein and Ehrenfest accepted that Planck's packets were not just an artifact of his calculation, but rather an indivisible unit of energy required to understand observations concerning the transfer of energy between atoms and the radiation field. Furthermore, Bohr developed a phenomenological model that envisaged spectral transitions between quantum stationary states and that finally explained the observed solar spectrum of hydrogen. Ever since, spectral observations have been interpreted in terms of quantum theory.

In 1923 de Broglie proposed a wave-particle theory of matter, which Schrödinger in 1926 recaste into the now familiar wave equation. Prior to this in 1925 Heisenberg and later with Pauli, Born, Jordan developed matrix mechanics. In 1926 Schrödinger and Eckart showed that matrix and wave mechanics were equivalent mathematical algorithms. In 1926 Born, Jordon and Dirac showed that the wave should be thought of in terms of describing the probability of finding a particle at a given point. In 1927 Heisenberg incorporated this uncertainty of position into his indeterminacy principle. Experimental verification of the wave character of matter came in 1927, when Davisson and Germer succeeded in diffracting electron beams in their passage through a crystal.

In 1927 Born and Oppenheimer showed that an approximate solution of the molecular wave equation could be obtained by first solving the electronic wave equation, with the nuclei in a fixed configuration and then solving a wave equation for the nuclei alone, which contains the electronic potential function. Provided that the potential function is known, then in principle all rovibrational energy levels may be calculated from the clamped-nucleus Schrödinger equation. In 1932 Dunham developed the connexion between the force constants in the potential energy function and the observed rovibrational spectrum of a diatomic molecule. Since then numerous analytical potential functions have been devised, with the potential function devised by Morse enjoying some success.

Studies of emission spectra of molecules in the visible and near ultraviolet began in the latter half of the 19th century. The regularity of the observed patterns aroused much interest and the term "band spectra" was attached to these patterns, since they were distinct from the sharp isolated lines of atomic spectra. In the latter half of the 19th century Deslanders established various empirical relations. However, infrared absorption of gases and liquids was developed in the early years of this century by Coblentz. The discovery of the Raman effect in 1928 provided a complementary investigative tool to infrared spectroscopy. After the second world war, the microwave region became accessible enabling the full spectral coverage for the rovibrational investigations of molecules. With the introduction of new technologies such as molecular beams, high power lasers and the Fourier-transform method, molecular rovibrational studies have been revitalized in the present decade in terms of both accuracy and precision.

Podolsky, Epstein and Dirac have shown how the correct quantum mechanical Hamiltonian can be obtained from a generalised coordinate system. Eckart developed the coordinate system for normal and non-linear molecules, which is suitable for the solution of the for small amplitudes of vibration. The system is based on molecule-fixed rotating coordinates and is designed to reduce the coupling terms with respect to the kinetic energy operator. Sayvetz generalised the theory so that it is applicable to molecules of all geometries. In this coordinate system, the first order terms of total angular momentum are small.

Group theory first entered physics with crystallography due to the work of mathematicians such as Weyl. The fundamental theorems of Wigner and Bargmann made the connexion between symmetries associated via group representations and the built-in linear structure of quantum theory. The utility of group theory is evident in the development of selection rules for rovibrational transitions and furthermore, in analyzing rovibrational spectra of molecules [3-4].The F and G matrix method of Wilson [7], which calculates force constants from experimental frequencies using an iterative approach, afforded a systematic and convenient means of incorporating symmetry properties of the molecule so as to achieve maximum factorization of the resulting secular determinant.

In 1947 Rees placed on solid footing the Rydberg-Klein-Rees (RKR) method. It yields equations which lead directly from the rotational and vibrational term values to the ordinates of the potential curve at a given energy. The RKR equations are usually employed to extract the potential curve from rovibrational spectra (usually without rotation).

Mainly due to the work of Nielsen and his coworkers contact transformations of the nuclear Schrödinger equation enabled the calculation of spectroscopic constants. Nielsen group transformed the nuclear Schrödinger equation into a new effective Hamiltonian which has the same eigenvalues, but different eigenfunctions, correct to a chosen order of magnitude. The contact transformation yields an expression for the rovibrational energy levels in the form of a convergent power series expansion in vibrational and rotational quantum numbers. The coefficients in the effective Hamiltonian are then the spectroscopic constants obtained from the analysis of rovibrational spectra.

In the case of diatomic molecules, the numerical integration devised by Cooley and Cashion enabled direct solutions of nuclear Schrödinger equation. Using this numerical approach, Kolos and Wolniewicz showed that given an accurate discrete ab initio potential energy surface for H_2, the calculated rovibrational energy levels are of very high accuracy, the error being limited only by the Born-Oppenheimer approximation itself.

For many decades spectroscopists viewed nuclear motion in terms of small amplitudes of vibrations and near rigid rotations about some vibrationally-averaged or equilibrium geometry (the latter being usually well defined by a potential energy surface). This traditional view has its roots in the formulation of a rovibrational Hamiltonian by Eckart [8], Wilson-Howard [9], Darling-Dennison [10] and Watson [11-12]. It basically contains six types of operators: vibrational kinetic operator, vibrational angular momentum operator, potential energy operator, a mass-dependent correction operator to the effective potential energy function, rotational kinetic energy operator and operators arising from the vibration-rotation interaction. It is this rovibrational Hamiltonian which has been successfully used in interpreting the rovibrational spectra of diatomic and triatomic molecules. The success in applying perturbation theory to the nuclear Hamiltonian in order to explain observed spectra has further entrenched this approach [13].

Watson's Hamiltonian [11-12] is the simplest quantum-mechanical form of Eckart's classical rovibrational Hamiltonian [8]. Basically, Watson's Hamiltonian differs from that previously used by experimentalists by a single operator (termed the Watson operator). As Watson has pointed out [11-12] a large amplitude of vibration which transforms a bent system (with 3N-6 degrees of freedom) into a linear system (with 3N-5 degrees of freedom) will necessarily cause the Watson term to become singular (since the formulation is inappropriate for linear systems). In order to circumvent the singularities various different approaches have been employed [14].

While highly accurate wavefunctions for diatomic molecules have become common place [15], prior to the 1970s no interplay existed between theory and experiment for even the simplest polyatomic molecule - the triatomic hydrogen cation, H_3^+. An ab initio variational calculation by Carney and Porter [16-18] of the vibrational band origins of H_3^+ (which preceded experiment [19-20]) gave credence to the notion that tractable polyatomic ab initio potential energy surfaces of spectroscopic quality could be constructed from first principles and therefore be of use to experimentalists. The concomitant decrease in the unit cost per computing power in the 1980s [21-22] facilitated rapid construction of a number of potential energy surfaces for spectroscopic analysis.

For H_3^+, the agreement between experiment and theory is impressive. Nevertheless, it is not yet all-embracing, since Carrington and coworkers [23-24] have observed some 27000 transitions from states near the dissociation limit in a small region (220 cm^{-1}) of the predissociation infrared spectrum. For high J transitions, the distortion and coupling constants for low J states do not provide a good enough guide to aid spectral assignment. It is therefore not surprising that Burton and von Nagy-Felsobuki [25] earmarked this two electron, three nuclei molecule as the benchmark molecule for computational chemistry in the 1990s.

Research on H_3^+ has provided a fertile test-bed for developing variational methods in determining accurate rovibrational transitions. In case of electronic structure calculations, the strategies for configuration calculations may differ, but the formulation of th electronic Hamiltonian is unwavering [21-22]. However, with respect to the nuclear Schrödinger equation not only are the solution strategies vastly different, but also the coordinate system in which the nuclear Hamiltonian is caste and the form of the nuclear Hamiltonian itself (i.e.whether the equilibrium geometry is embedded within it or not). That the situation is not yet settled is evident by the recent article of Sutcliffe [26], who has detailed the use of perimetric coordinates in the rovibrational Hamiltonian for future calculations of the rovibrational spectra of triatomic molecules.

The purpose of this monograph is to detail an ab initio variational method of calculating the rovibrational spectra of triatomic molecules based on the Eckart-Watson Hamiltonian. Quantal derivative methods which rely on inversion of rovibrational spectra to deduce intramolecular potential surfaces as well as perturbative procedures are outside the scope of this monograph. The review of Carney, Sprandel and Kern [27] in 1977, the potential-energy surfaces and reaction dynamics issue of the Journal of the Chemical Society in 1987 [28], the proceedings of Faraday symposium 23 on molecular vibrations in 1988 [29] and the review of Searles and von Nagy-Felsobuki [14] in 1991, all fall within the general theme of this monograph. However, in order to place in context the nuclear Hamiltonian and solution methodology detailed in this monograph, alternative Hamiltonians and solution strategies shall be sketched below. These will be shown to yield results, where comparisons can be made, that are in excellent agreement with the Eckart-Watson Hamiltonian and the solution algorithm detailed here.

§.2. Internal Coordinate Hamiltonian.

The Watson Hamiltonian for non-linear molecules becomes singular when the molecule becomes linear. Hence this Hamiltonian is not appropriate for molecules undergoing large amplitudes of vibration (i.e. floppy molecules). One approach is to reject the normal coordinate scheme altogether and develop a coordinate system such that the Hamiltonian does not become singular when the molecule becomes linear. Such a coordinate system is the internal coordinates consisting of two bond lengths and the included angle.

Sutcliffe [30] and Carter et al. [31] have derived a full rovibrational Hamiltonian for any triatomic molecule ABC, in terms of the internal vibrational coordinate system: R_1, R_2 and θ. It should be noted that the Eckart frame is not used in this model. Their vibrational Hamiltonian is identical to that obtained by Lai [32] and Hagstrom, who based their derivation on the Wilson-Howard Hamiltonian. Handy [33] using a computer algebra programme REDUCE has rederived this Hamiltonian. The rovibrational Hamiltonian has form,

$$\hat{H}(R_1, R_2, \theta, \alpha, \beta, \gamma) = \hat{H}_V(R_1, R_2, \theta) + \hat{H}_{VR}(R_1, R_2, \theta, \alpha, \beta, \gamma) + \hat{V} \qquad (2.1)$$

where the vibrational Hamiltonian is given by,

$$\hat{H}_V(R_1, R_2, \theta) = -\frac{\hbar^2}{2}\left\{\frac{2\cos\theta}{m_B R_1 R_2} - \frac{2\cos\theta}{m_B R_2}\frac{\partial}{\partial R_1} - \frac{2\cos\theta}{m_B R_1}\frac{\partial}{\partial R_2}\right.$$

$$+\frac{1}{\mu_1}\frac{\partial^2}{\partial R_1^2} + \frac{1}{\mu_2}\frac{\partial^2}{\partial R_2^2} + \frac{2\cos\theta}{m_B}\frac{\partial^2}{\partial R_1 \partial R_2}$$

$$+\left[\cot\theta\left(\frac{1}{m_1 R_1^2} + \frac{1}{m_2 R_2^2}\right) + \frac{4\sin\theta - 2\,\mathrm{cosec}\,\theta}{m_B R_1 R_2}\right]\frac{\partial}{\partial\theta}$$

$$-\frac{2\sin\theta}{m_B R_2}\frac{\partial^2}{\partial\theta\partial R_1} - \frac{2\sin\theta}{m_B R_1}\frac{\partial^2}{\partial\theta\partial R_2}$$

$$\left.+\left[-\frac{2\cos\theta}{m_B R_1 R_2} + \frac{1}{\mu_1 R_1^2} + \frac{1}{m_2 R_2^2}\right]\frac{\partial^2}{\partial\theta^2}\right\} \tag{2.2}$$

The rotational and the rovibrational coupling terms are given by,

$$\hat{H}_{VR}(R_1, R_2, \theta, \alpha, \beta, \gamma) = i\hbar\left\{\frac{\sin\theta}{m_B R_1}\frac{\partial}{\partial R_2} + \left[\frac{\cos\theta}{m_B R_1 R_2} - \frac{1}{m_1 R_1^2}\right]\frac{\partial}{\partial\theta}\right.$$

$$\left.-\frac{1}{2}\left[\frac{\cot\theta}{m_1 R_1^2} + \frac{2\sin\theta - \mathrm{cosec}\,\theta}{m_B R_1 R_2}\right]\right\}\hat{\Pi}_y + \frac{1}{2\mu_1 R_1^2}(\hat{\Pi}_x^2 + \hat{\Pi}_y^2)$$

$$-\frac{1}{2}\left[\frac{2\cot\theta\,\mathrm{cosec}\,\theta}{m_B R_1 R_2} - \mathrm{cosec}^2\theta\left(\frac{1}{\mu_1 R_1^2} + \frac{1}{\mu_2 R_2^2}\right) + \frac{1}{\mu_1 R_1^2}\right]\hat{\Pi}_z^2$$

$$-\frac{1}{2}\left[\frac{\cot\theta}{m_1 R_1^2} - \frac{\mathrm{cosec}\,\theta}{m_B R_1 R_2}\right](\hat{\Pi}_x\hat{\Pi}_y + \hat{\Pi}_y\hat{\Pi}_x) \tag{2.3}$$

The rotational operators are given in terms of the Euler angles (α, β, γ) as,

$$\hat{\Pi}_x = -i\hbar\cos\gamma\cot\beta\frac{\partial}{\partial\gamma} - i\hbar\sin\gamma\frac{\partial}{\partial\beta} + i\hbar\cos\gamma\,\mathrm{cosec}\,\beta\frac{\partial}{\partial\alpha}$$

$$\hat{\Pi}_y = i\hbar\sin\gamma\cot\beta\frac{\partial}{\partial\gamma} - i\hbar\cos\gamma\frac{\partial}{\partial\beta} + i\hbar\sin\gamma\,\mathrm{cosec}\,\beta\frac{\partial}{\partial\alpha} \tag{2.4}$$

$$\hat{\Pi}_z = -i\hbar \frac{\partial}{\partial \gamma}$$

and μ_1 and μ_2 are given by,

$$\mu_1 = \left\{ \frac{1}{m_A} + \frac{1}{m_B} \right\}^{-1}$$

(2.5)

$$\mu_2 = \left\{ \frac{1}{m_C} + \frac{1}{m_B} \right\}^{-1}$$

Lai [32] used the Hamiltonian (given by equation (2.2)) to calculate vibrational band origins of the water molecule. The Hamiltonian was initially separated into three one-dimensional problems, which were solved numerically using the Numerov method [34]. The resulting eigenfunctions were used as a basis for the variational solution of the full vibrational problem. Lai [32] showed that the solutions obtained using this variational technique are more accurate than those calculated using perturbation theory.

Cropek and Carney [35] developed a solution algorithm based on that of Lai [32] and Hagstrom. Carney and coworkers used this algorithm to calculate vibrational band origins of H_2O, O_3 and isotopomers of H_3^+ [35]. Table 1.1 compares the vibrational eigenenergies for O_3 using the Lai [32] and Hagstrom model with eigenenergies obtained from t coordinate Hamiltonian of Carney et al. [36]. The vibrational eigenenergies are in excellent agreement, yielding a standard deviation of only 1.99 cm^{-1}. Table 1.1 further illustrates that the t coordinate Hamiltonian is appropriate for small amplitudes of vibration of bent triatomic molecules.

In the case of H_2O the results obtained using the Lai and Hagstrom model are in better agreement with the experimental eigenenergies than those calculated using the t coordinate approach [37]. However, for the H_3^+ isotopomers, the t coordinate scheme is found to be more efficient except in the case of H_2D^+ [38]. Carney et al. [38] also compared the variational results for H_3^+ with those calculated using a second order perturbation algorithm. The variational results are more accurate when compared to experiment.

Table 1.1 Comparison of Vibrational Eigenenergies using the **t** coordinate Hamiltonian and Internal Coordinate Hamiltonian for Ozone[a].

$v_1v_2v_3$	Sym	t Coordinate[b]	Internal Coordinate[c]
000	A	1454.8	1454.7
010	A	700.0	701.6
001	B	1041.2	1041.0
100	A	1101.2	1099.3
020	A	1395.9	1399.4
011	B	1721.5	1722.9
110	A	1791.0	1790.5
002	A	2054.9	2054.2
030	A	2087.0	2090.8
101	B	2106.1	2103.8
200	A	2198.3	2195.1
021	B	2397.0	2398.1
120	A	2480.5	2480.4
012	A	2717.6	2715.8

a) Averaged over degenerate modes.
b) See reference [36].
c) See reference [37].

§.3. Generalised. Body-fixed Coordinate Hamiltonian.

Sutcliffe, Tennyson and coworkers [39-41] developed a rovibrational Hamiltonian in terms of a body-fixed scattering coordinate system which is appropriate for the study of molecules with small and large amplitude vibrations. In the body-fixed scattering coordinate system, the ABC molecule is considered as the interaction of a diatomic molecule, AC and a nucleus, B. The body-fixed coordinates (r, R, θ) are defined as the internuclear distance of the diatomic AB, the distance from the diatomic centre-of-mass to the nucleus B and the angle between r and R, respectively. The rotational motion is separated from the vibrational motion by embedding a body-fixed coordinate frame. Since a non-rigid system is being discussed, the definition of an equilibrium structure may not be possible, and the body-fixed Eckart frame (which requires definition of an equilibrium structure) is therefore not suitable. Sutcliffe, Tennyson and coworkers [39-42] use either a frame where the z-axis is fixed along the R coordinate or along r. Although this choice is arbitrary in the full be rovibrational problem, it should be made so as to minimise the Coriolis interaction terms if the vibration-only problem is to be considered. The x-axis is placed in the plane of the molecule and the y-axis is normal to the plane and forms a right-handed system.

Sutcliffe, Tennyson and coworkers [43-45] have methodically examined the embedded axes in the nuclear Hamiltonian formulation. Sutcliffe and Tennyson [46] have shown that one is free to choose any coordinate system and set of axes and then transform the full nuclear Hamiltonian into appropriate body-fixed form. The choice of Hamiltonian is reduced to the definition of a single parameter to define the coordinate system and another to define the embedding of the axes. Sutcliffe [43] has produced a procedure for embedding an arbitrary set of internal axes.

In the formulation, Sutcliffe and Tennyson [44, 47] have assumed that the internal coordinates (free of translation) for a triatomic molecule, t_j are related to the laboratory-fixed coordinates x_i by the transformations,

$$t_j = \sum_{i=1}^{3} x_i L_{ij} \tag{3.1}$$

where,

$$L = \begin{bmatrix} 0 & 1 \\ -1 & -g \\ 1 & g-1 \end{bmatrix} \tag{3.2}$$

and $0 \leq g \leq 1$. The bond length vector from nucleus B to nucleus C is t_1. The coordinate system is selected by specifying the value of g. For example, the internal coordinate system of section §.2. is given by setting g=1, so that t_2 is the bond length vector from particle B to particle A. The scattering coordinates are obtained by setting,

$$g = \left\{ \frac{m_B}{m_B + m_C} \right\}^{-1} \tag{3.3}$$

so that t_2 becomes the vector from the centre-of-mass of the diatomic system B-C to particle A.

The translation-free internal coordinates (t_i) are now transformed by an orthogonal transformation (given by C) so that,

$$t_i = Cz_i \tag{3.4}$$

where the matrix C defines the three Euler angles (α, β, γ) which are required to rotate the laboratory fixed axes into the body-fixed frame and the coordinates z_i are functions of the three vibrational coordinates q_k. The body-fixed Hamiltonian can then be written in terms of the angular momentum operators and of operators and functions of the vibrational coordinates, q_k.

The embedding of the body-fixed axes is determined by specification of the matrix C. For example, in order that the embedded z-axis lies along t_1, t_2 is in the positive x half of the xz plane, and the three body-fixed axes form a right-handed set, then,

$$C^T t_1 = r \begin{bmatrix} 0 \\ 0 \\ 1 \end{bmatrix}$$

$$\tag{3.5}$$

$$C^T t_2 = R \begin{bmatrix} \sin\theta \\ 0 \\ \cos\theta \end{bmatrix}$$

where r and R are the lengths of t_1 and t_2 respectively, q is the angle between t_1 and t_2 in a right-handed system.

If the body-fixed Hamiltonian is then allowed to operate on the manifold of rotational functions $|Jkm\rangle$ and the results multiplied from the left by $\langle Jk'm|$, taking integration over the rotational variables the resulting effective body-fixed Hamiltonian has form,

$$\hat{H} = \hat{T}_V^1 + \hat{T}_V^2 + \hat{T}_{VR}^1 + \hat{T}_{VR}^2 + \hat{V} \tag{3.6}$$

where $\hat{T}_V^1$ and $\hat{T}_V^2$ are the kinetic energy operators associated with the vibration-only problem, whereas $\hat{T}_{VR}^1$ and $\hat{T}_{VR}^2$ are the kinetic energy operators which couple vibration and rotation. It should be noted that if scattering coordinates are chosen then $\hat{T}_V^2$ and $\hat{T}_{VR}^2$ are null operators and the Hamiltonian reduces to one used by the authors in their previous work [48]. On the other hand, if the bond lengths and included bond angle are chosen then the Hamiltonian given by equations (2.2) and (2.3) collapses to that given by Lai [32] and Hagstrom.

In the scattering coordinate approach of Tennyson and Sutcliffe [39-48] the kinetic energy operator takes on the form,

$$\hat{T} = \hat{T}_V^1 + \hat{T}_{VR}^1 \tag{3.7}$$

where,

$$\hat{T}_V^1 = \delta_{j'j}\delta_{k'k}\left[-\frac{\hbar^2}{2\mu_1}\frac{\partial^2}{\partial r^2} - \frac{\hbar}{2\mu_2}\frac{\partial^2}{\partial R^2} + \frac{\hbar^2}{2}j(j+1)\left(\frac{1}{\mu_1 r^2} + \frac{1}{\mu_2 R^2}\right)\right] \tag{3.8}$$

and,

$$\hat{T}_{VR}^1 = \delta_{j'j}\delta_{k'k}\frac{\hbar^2}{2\mu_1 r^2}(J(J+1) - 2k^2) - \delta_{j'j}\frac{\hbar^2}{2\mu_1^2 r_1^2}(\delta_{k'k+1}C_{Jk}^+C_{jk}^+ + \delta_{k'k-1}C_{Jk}^-C_{jk}^-) \tag{3.9}$$

where $C_{Jk}^{\pm} = (J(J+1) - k(k\pm1))^{1/2}$ and so are the usual step up and step down coefficients and,

$$\mu_1 = \left\{\frac{1}{m_B} + \frac{1}{m_C}\right\}^{-1}$$

$$\mu_2 = \left\{\frac{1}{m_A} + \frac{g^2}{m_B} + \frac{(1-g)^2}{m_C}\right\}^{-1} \tag{3.10}$$

By choosing a basis of products of the form,

$$\varphi_{nkj}^J(r, R)\,\Theta_{jk}(\theta) \tag{3.11}$$

where Θ_{jk} are the normalised associated Legendre functions and φ_{nkj}^J are products of the central-field functions. It can be shown that the $\operatorname{cosec}^2\theta$ terms in $\hat{T}_{VR}^1$ cancel with those which arise from the operation of the derivative operators in $\hat{T}_V^1$ and $\hat{T}_{VR}^2$ upon the φ_{nkj}^J, thereby removing the incipient singularities.

The effective kinetic energy operator works only on functions that contain radial variables. Tennyson and Sutcliffe have found that of the radial functions tested [39-48], the Morse oscillator-like functions (denoted as $|v\rangle$) [48] have proven to be both flexible and computationally tractable in multi-dimensional calculations. The flexibility arises from the Morse parameters, which are usually treated as variational parameters and so are optimised accordingly [44, 49]. Hence, in the scattering coordinate scheme only two types of kinetic energy operators arise, namely: r^2 and $\partial^2/\partial r^2$. They have derived analytic closed forms for matrix elements of the second order differential operator [48]. Integrals involving the r^2 operator are evaluated using Gauss-Laguerre integration.

In order to increase the rate of convergence of high-lying vibrational levels (with respect to basis set size) Bačić et al. [50-53] have developed a solution algorithm which differs from previously used methods in the choice of basis functions for the expansion of the wavefunctions. The traditional methods generally employ products of one-dimensional single-centre basis functions. For example, in the scattering coordinates of Tennyson et al. [39-41, 49] the basis set which is normally used consists of products of the form,

$$F_1(r)F_2(R)\Theta_{jk}(\theta)D_{Mk}^{J}(\alpha,\beta,\gamma) \hspace{3cm} (3.12)$$

where $F_1(r)$ and $F_2(R)$ represent one-dimensional orthogonal functions of the radial coordinates, $\Theta_{jk}(\theta)$ is the associated Legendre polynomial, and $D_{Mk}^{J}(\alpha,\beta,\gamma)$ is the rotational wavefunction. Since wavefunctions corresponding to high-lying levels may be extensively delocalised, a set of oscillator basis functions centred at some point on the surface cannot provide a compact, rapidly converging representation of a delocalised, anharmonic vibrational wavefunction.

Bačić and Light [51] have detail the desirable properties of basis functions, concluding that single-centre oscillator basis functions are not suitable. They combine the discrete variable representation (DVR) of the angular, bend coordinate with the real, distributed Gaussian basis (DGB) for the radial parts of the vibrational wavefunctions [50-51]. The DVR is used for the representation of the bend angle and a separate distributed Gaussian basis is used at each angle to represent motion in the other possible degrees of freedom.

The DVR-DGB approach has be used to study the LiCN/LiNC system which is extremely floppy [51-53]. This system has been studied by Farantos and Tennyson [54] using a direct product, single-centred basis set. Comparison of the vibrational energy levels using the two approaches indicates that the DVR-DGB has faster convergence with respect to basis size. Bačić et al. [51-53] use the scattering coordinate system and Hamiltonian of Tennyson et al. [39-41, 49]. However, it should be noted that the basis set constructed is applicable to any coordinate system.

Carter and Meyer [55] have shown the pitfalls of the DVR method with indiscriminate use. The DVR approach can yield converged results which violate the variational principle. Consequently, Tennyson's group maintain two sets of code for the general rovibrational problem: the DVR code for high vibration-rotation excited states and the finite basis representation (given by equation (2.11)) for low-lying states.

For small amplitudes of vibration it would be anticipated that agreement between the solutions of the scattering coordinate and normal coordinate Hamiltonians would be good. Henderson, Miller and Tennyson (HMT) [56] (using the Searles, Dunne and von Nagy-Felsobuki (SDF) electronic and dipole moment surface [57-58]) calculated the rovibrational energy levels ($J \leq 4$) and transition intensities of $^7Li_3^+$ and $^7Li_2{}^6Li^+$. Their calculations corroborated the earlier calculation of Searles et al. [58] which predicted band origins for $^7Li_3^+$, but extended that work by calculating the rotational constants and rovibrational wave functions of the two most abundant isotopomers of Li_3^+.

Table 1.2 compares the results obtained by Searles and von Nagy-Felsobuki [59] using the t coordinate and s coordinate vibrational Hamiltonian (see Chapter VI), with those of HMT for $^7Li_3^+$ and $^7Li_2{}^6Li^+$. It should be noted that the only common feature between both solution algorithms is the SDF force field. Clearly both methods yield the same assignment for the first ten vibrational band origins of both isotopomers and moreover, are in accord to within 0.03 cm^{-1} for the vibration-only ($J=0$) problem. Furthermore, these results suggest that the use of a third-order perturbation expression for the Watson operator in t and s coordinates is appropriate (see Chapter VI). Therefore it is apparent that at this level of sophistication both formulations have in fact converged.

§.4. Rectilinear Vibrational Displacement Coordinates within the Eckart Framework.

In order to avoid problems when large displacements of the bending coordinate are encountered, Hougen et al. [60] introduced a coordinate system which is based on two bond length displacements, Δr_{12} and Δr_{23} and a bending coordinate, $\Delta \gamma$. By setting $\Delta r_{12} = \Delta r_{23} = 0$, a rovibrational Hamiltonian within the Eckart framework was developed. The resulting Hamiltonian does not constrain the molecule to small amplitude bending vibrations. This work formed the basis of the non-rigid-bender Hamiltonian, which has been discussed in the review of Bunker [61].

Jensen [62-64] has developed a rovibrational Hamiltonian for a triatomic molecule based on three curvilinear stretching coordinates Δr_i and expanded in the form of a power series with a Morse expansion variable,

$$y_i = 1 - exp(-a_i \, \Delta r_i) \tag{4.1}$$

Table 1.2 Calculated Vibrational Band Origins of the Isotopomers of Li_3^+ (in cm^{-1})[a].

$v_1v_2v_3$	$^7Li_3^+$			$^7Li_2{}^6Li^+$		
	Sym	SDF[b]	HMT[c]	Sym	SDF[d]	HMT[c]
010	E	225.98	225.98[a]	A_1	231.40	231.40
001	E	225.98	225.98[a]	B_2	231.92	231.93
100	A_1	298.76	298.77	A_1	307.29	307.30
020	A_1	447.76	447.80	A_1	458.76	458.80
002	E	451.05	451.09[a]	A_1	462.04	462.06
011	E	451.05	451.09[a]	B_2	462.80	462.81
101	E	520.54	520.56[a]	B_2	534.27	534.28
110	E	520.54	520.56[a]	A_1	534.84	534.86
200	A_1	595.47	595.48	A_1	612.33	612.34
Zero-Point Energies		380.872			391.141	

a) Averaged over degenerate modes. Reproduced with permission from reference [59].
b) See reference [58].
c) See reference [56].
d) See reference [59].

where a_i are the molecular constants obtained from the potential energy function. This coordinate system is most suitable for study of molecules whose vibrations are dominated by bond stretches. The eigenvalues are obtained variationally using Morse variable as a basis for each bond. The rovibrational Hamiltonian for this system is given by,

$$\hat{H} = \frac{1}{2}\sum_\alpha \sum_\beta (\hat{\Pi}_\alpha - \hat{\pi}_\alpha)\, \mu_{\alpha\beta}(\Delta r_1, \Delta r_2, \Delta r_3)\, (\hat{\Pi}_\beta - \hat{\pi}_\beta) + U(\Delta r_1, \Delta r_2, \Delta r_3)$$

$$+ \frac{1}{2}\sum_{j=1}^{3}\sum_{k=1}^{3} \hat{P}_j G_{jk}(\Delta r_1, \Delta r_2, \Delta r_3)\, \hat{P}_k + V(\Delta r_1, \Delta r_2, \Delta r_3) \qquad (4.2)$$

where $\alpha, \beta = x, y, z$. $\hat{\Pi}_\alpha$ are the components of the total angular momentum measured along the molecule-fixed axes, $\mu_{\alpha\beta}$ is the inverse effective moment of inertia tensor, $\hat{P}_j$ are the momenta conjugate to Δr_j, G_{jk} are the usual G matrix elements and U is a pseudopotential term which originates from the conversion from the classical to the quantum mechanical Hamiltonian.

The analytical form of the potential energy function is represented by,

$$V(y_1, y_2, y_3) = \frac{1}{2}\sum_i^3\sum_j^3 f_{ij}y_i y_j + \frac{1}{6}\sum_i^3\sum_j^3\sum_k^3 f_{ijk}y_i y_j y_k$$

$$+ \frac{1}{24}\sum_i^3\sum_j^3\sum_k^3\sum_l^3 f_{ijkl}y_i y_j y_k y_l \qquad (4.3)$$

where the $f_{jkm...}$ are the expansion coefficients.

Jensen et al. [65] least squares fitted equation (4.3) to 69 of the ab initio points generated by Burton et al. [66]. They omitted all ab initio points more than $13000\ cm^{-1}$ above the dissociation energy of H_3^+; the total energy range covered in the fitting is $44500\ cm^{-1}$. Moreover, the root-mean-square deviation (r.m.s) of their fit is $46\ cm^{-1}$.

Table 1.3 compares the calculated vibrational band origins of H_2D^+ and HD_2^+ calculated by Tennyson and Sutcliffe [67] and Miller and Tennyson [68] using a scattering coordinate model with the model of Jensen et al. [65]. Although both models use the ab initio surface of Burton et al. [66], the vibrational band origins differ due to the significant difference in the analytical potentials and variational methods; furthermore, the model of Jensen et al. [65] uses a vibrational Hamiltonian truncated to fourth-order in y_i, whereas Miller and Tennyson [68] use the full vibrational Hamiltonian. Of course all methods may suffer from a basis set incompleteness problem. For H_2D^+ the differences between the two approaches are within $20\ cm^{-1}$, whereas for HD_2^+ within $14\ cm^{-1}$.

Table 1.3 Comparison Calculated Vibrational Band Origins of the Isotopomers of H_3^+ (in cm^{-1}).

$\nu_1\nu_2\nu_3$	Method 1[a]	Method 2[b]
	H_2D^+	
010	2203	2200
001	2332	2325
100 [c]	2989	2984
020	4282	4286
011	4459	4446
002	4602	4581
110	5035	5028
101	5239	5220
	D_2H^+	
010	1965	1959
001	2075	2073
100 [d]	2733	2728
020	3815	3809
011	4038	4037
002	4059	4042
110	4644	4630
101	4668	4660
200	5380	5366

a) See reference [67-68].

b) See reference [65].

c) Experimental value 2992.488 ± 0.016 cm^{-1}. See reference [69].

d) Experimental value 2736.997 ± 0.026 cm^{-1}. See reference [70].

Comparison with experiment [69-70] is possible for the vibrational band origins. For (H_2D^+ and HD_2^+) the models of Miller and Tennyson [68] and Jensen et al. [65] are within (3 cm^{-1}, 8 cm^{-1}) and (4 cm^{-1}, 9 cm^{-1}) of the experimental values respectively.

In order to study molecules for which the bending vibration of the molecule is important, Jensen [63] developed the MORBID Hamiltonian. The coordinate system used is the (Δr_{12}, Δr_{23}, γ) system defined by Hougen et al. [60]. The rovibrational Hamiltonian is expanded to quartic terms in the Morse-oscillator expansion given by equation (4.3). The eigenvalues of this Hamiltonian are obtained variationally using products of Morse oscillator stretching functions and numerical bending functions as a basis.

The analytical form of the potential energy function is represented by,

$$V(\Delta r_{12}, \Delta r_{23}, \overline{\gamma}) = V_o(\overline{\gamma}) + \sum_i^3 F_i(\overline{\gamma}) \, y_i + \sum_{i \leq j} F_{ij}(\overline{\gamma}) \, y_i y_j$$

$$+ \sum_{i \leq j \leq k} F_{ijk}(\overline{\gamma}) \, y_i y_j y_k + \sum_{i \leq j \leq k \leq l} F_{ijkl}(\overline{\gamma}) \, y_i y_j y_k y_l \qquad (4.4)$$

where the indices i, j, k, ... assume the values of 1, 2 or 3. The expansion coefficients, $F_{ijk...}$, depend on the instantaneous value of the bond angle supplement ($\overline{\gamma} = \pi - \gamma$). The expansion coefficients are given by,

$$V_o(\overline{\gamma}) = \sum_{i=2}^8 f_o{}^i (\cos \gamma_e - \cos \overline{\gamma})^i$$

$$F_j(\overline{\gamma}) = \sum_{i=1}^3 f_j{}^{(i)} (\cos \gamma_e - \cos \overline{\gamma})^i \qquad (4.5)$$

$$F_{jk...}(\overline{\gamma}) = f_{jk...}{}^0 + \sum_{i=1}^N f_{jk...}{}^{(i)} (\cos \gamma_e - \cos \overline{\gamma})^i$$

where γ_e is the equilibrium value of γ.

In order to calculate ab initio vibrational band origins, the values of the input data $r_{j2}{}^e$, a_j, γ_e and $f_{jkm...}{}^{(i)}$ are obtained by fitting the expansion of equation (4.5) through the discrete ab initio potential energy surface.

The quantum mechanical Hamiltonian is of the form [62-64],

$$\hat{H} = \frac{1}{2} \sum_\alpha \sum_\beta (\hat{\Pi}_\alpha - \hat{\pi}_\alpha) \, \mu_{\alpha\beta} \, (\Delta r_{12}, \Delta r_{23}, \gamma) \, (\hat{\Pi}_\beta - \hat{\pi}_\beta)$$

$$+ \frac{1}{2} \sum_{j=1}^{3} \sum_{k=1}^{3} \hat{P}_j \, G_{jk}^{(r)}(\Delta r_{12}, \Delta r_{23}, \gamma) \, \hat{P}_k + U_1(\Delta r_{12}, \Delta r_{23}, \gamma)$$

$$+ U_0(\gamma) + V(\Delta r_{12}, \Delta r_{23}, \gamma) \hspace{2cm} (4.6)$$

where α, β = x, y, z, γ. $\hat{\Pi}_\alpha$, α = x, y, z, are the total angular momentum operators along the molecule fixed axis with $\hat{\Pi}_\gamma$ being $-i\hbar\partial/\partial\gamma$; $\mu_{\alpha\beta}$ is the inverse of the effective inertial tensor; $\hat{\pi}_\alpha$ are the vibrational angular momentum operators; $\hat{P}_1$ and $\hat{P}_3$ are the momenta conjugate to Δr_{12} and Δr_{23}; U_1 and U_0 are pseudopotential terms obtained from converting the classical Hamiltonian equation to quantum mechanical form and V is the potential function given by equation (4.5).

The $G_{jk}^{(r)}$ matrix elements can be expressed as a power series in the Δr_{j2} with γ-dependent coefficients and so can be expressed as a power series in y_j [62-64]. Further, the $\mu_{\alpha\beta}$ matrix elements can also be expressed as a power series in y_1 and y_3 with γ-dependent coefficients [62-64]. Since matrix elements which involve V (see equation (4.5)), U and $\hat{\pi}$ operators can also be expressed in terms of a power series in y_j with γ-dependent coefficients, the total Hamiltonian can therefore be expanded to the fourth order in the stretching coordinates y_j and $\hat{P}_j$, with expansion coefficients generally being functions of γ and $\hat{\Pi}_\gamma$. Jensen uses REDUCE II [71] in order to to transform these functions as polynomials in y_j with the γ-dependent expansion coefficients.

In order to construct triple product basis functions $|v_1 v_2 v_3 \rangle$, Jensen [62-64] initially diagonalises a "pure" stretching Hamiltonian spanned by a symmeterised basis of Morse oscillator functions (denoted as v_1 and v_3) using the closed expressions for the various matrix elements as given by Špirko et al. [65]. The bending basis functions (denoted as v_2) are obtained from the diagonalisation of the "pure" bending Hamiltonian. The bending Hamiltonian is that for a molecule bending and rotating about the z axis at fixed Δr_{j2} [72]. The bending functions are numerical since the Hamiltonian is integrated using the Numerov-Cooley method. All the matrix elements in setting up the triple product basis are stored on the disk and are retrieved for the diagonalisation of the total Hamiltonian matrix.

Jensen and coworkers have used the MORBID Hamiltonian to study a wide range of triatomic molecules. In general, good agreement with the available experimental data is found.

REFERENCES TO CHAPTER I

1 Harvey NE (1957) A history of luminescence, American Philosophical Society, Baltimore

2 Mehra J, Rechenberg H (1982) The historical development of quantum theory, Vols 1-4, Springer-Verlag, Berlin

3 Herzberg G (1945) Molecular Spectra and Molecular structure, Vols 1-3, Van Nostrand Reinhold, Melbourne

4 Huber KP, Herzberg G (1979) Constants of diatomic molecules, Vol. 4, Van Nostrand Reinhold, Melbourne

5 Robinson R (1966) Aboriginal myths and legends, Sun Books, Melbourne

6 Bynnum WF, Browne EJ, Porter R (1969) Dictionary of the history of science, Princeton University Press, New Jersey

7 Wison EB, Decius JC, Cross PC (1955) Molecular vibrations, Dover Publications, New York

8 Eckart C (1935) Phys Rev 47:552

9 Wilson EB, Howard JB (1936) J Chem Phys 4:260

10 Darling BT, Dennison DM (1940) Phys Rev 57:128

11 Watson JKG (1968) Mol Phys 15:479

12 Watson JKG (1970) Mol Phys 19:465

13 Papoušek D, Aliev MR (1982) Molecular vibrational-rotational spectra, Elsevier, Amsterdam

14 Searles DJ, von Nagy-Felsobuki EI (1991) In: Vibrational structure and spectra, Durig JR (Ed), Vol 19, Elsevier, New York

15 Schmidt-Mink I, Muller W and Meyer W (1985) Chem Phys 92:263

16 Carney GD, Porter RN (1974) J Chem Phys 60:4251

17 Carney GD, Porter RN (1976) J Chem Phys 65:3547

18 Carney GD, Porter RN (1980) Phys Rev Letts 45:537

19 Oka T (1980) Phys Rev Lett 45:531

20 Shy JT, Farely JW, Lamb WE, Wing WH (1980) Phys Rev Lett 45:535

21 Dykstra CE (Ed) (1984) Advanced theories and computational approaches to the electronic structure of molecules, Reidel, New York

22 Lagana A (Ed) (1989) Supercomputer algorithms for reactivity, dynamics and kinetics of small molecules, Kluwer Academic Publishers, Boston

23 Carrington A, Buttenshaw J, Kennedy RA (1982) Mol Phys 45:753

24 Carrington A, Kennedy RA (1984) J Chem Phys 81:91

25 Burton PG, von Nagy-Felsobuki EI (1988) Chem Aust 55:408

26 Sutcliffe BT (1992) Mol Phys 75:1233

27 Carney GD, Sprandel LL, Kern CW (1978), Adv Chem Phys 37:305

28 Handy NC, Gaw JF, Simandiras ED (1987) J Chem Soc Faraday Trans 2 83:1577

29 Jensen P (1988) J Chem Soc Faraday Trans 2 84:1315

30 Sutcliffe BT (1983) Mol Phys 48:561

31 Carter S, Handy NC, Sutcliffe BT (1983) Mol Phys 49:745

32 Lai EKC (1975) Masters Thesis, Indiana Unitversity, Bloomington, Indiana

33 Handy NC (1987) Mol Phys 61:207

34 Blatt JM (1967) J Comput Phys 1:382

35 Cropek D, Carney GD (1984) J Chem Phys 80:4280

36 Carney GD, Langhoff SR, Curtiss LA (1977) J Chem Phys 66:3724

37 Carney GD, Curtiss LA, Langhoff SR (1976) J Mol Spectrosc 61:371

38 Carney GD, Adler-Golden SM, Lesseski DC (1986) J Chem Phys 84:3921

39 Tennyson J, Sutcliffe BT (1983) J Mol Spectrosc 101:71

40 Tennyson J, Sutcliffe BT (1983) J Chem Phys 79:43

41 Tennyson J, Sutcliffe BT (1984) Mol Phys 51:887

42 Tennyson J, Miller S, Henderson JR (1990) J Chem Soc Faraday Trans 2 86:1963

43 Sutcliffe BT (1982) Current aspects of quantum chemistry, Carbó R (Ed) Elsevier, Amsterdam

44 Sutcliffe BT, Tennyson J (1986) Mol Phys 58:1053

45 Sutcliffe BT, Tennyson J (1987) Molecules in physics, chemistry and biology, Csizmadia I, Maruani J (Eds), Reidal, Dordrecht

46 Sutcliffe BT, Tennyson J (1987) J Chem Soc Faraday Trans 2 83:1663

47 Sutcliffe BT, Tennyson J (1988) Molecules in physics, chemistry and biology, Maruani J (Ed) Vol.12, Kluwer Academic Publishers, Boston

48 Tennyson J, van der Avoird A (1982) J Chem Phys 76:5710

49 Tennyson J (1985) Comput Phys Commun 38:39

50 Bačić Z, Light JC (1989) Ann Rev Phys Chem 40:469

51 Bačić Z, Light JC (1986) J Chem Phys 85:4594

52 Light JC, Bačić Z (1987) J Chem Phys 87:4008

53 Bačić Z, Whitnell RM, Brown D, Light JC (1988) Comp Phys Commun 51:35

54 Farantos SC, Tennyson J (1985) J Chem Phys 82:800

55 Carter S, Meyer W (1990) J Chem Phys 93:8902

56 Henderson JR, Miller S, Tennyson J (1988) Spectrochim Acta A44:1287

57 Dunne SJ, Searles DJ, von Nagy-Felsobuki EI (1987) Spectrochim Acta A43:699

58 Searles DJ, Dunne SJ, von Nagy-Felsobuki EI (1988) Spectrochim Acta A44:505

59 Searles DJ, von Nagy-Felsobuki EI (1989) Aust J Chem 42:737

60 Hougen JT, Bunker PR, Johns JWC (1970) J Mol Spectrosc 34:136

61 Bunker PR (1983) Ann Rev Phys Chem 34:59

62 Jensen P (1988) J Chem Soc Faraday Trans 2 84:1315

63 Jensen P (1988) J Mol Spectrosc 128:478

64 Jensen P (1989) J Mol Spectrosc 133:438

65 Jensen P, Špirko V, Bunker PR (1986) J Mol Spectrosc 115:269

66 Burton PG, von Nagy-Felsobuki EI, Doherty G, Hamilton M (1985) Mol Phys 55:527

67 Tennyson J, Sutcliffe BT (1985) Mol Phys 56:1175

68 Miller S, Tennyson J (1987) J Mol Spectrosc 126:183

69 Špirko V, Jensen P, Bunker PR, Čejchan A (1985), J Mol Spectrosc 112:183

70 Lubic KG, Amano T (1984) Can J Phys 62:1886

71 Hearn AC (1985) REDUCE User's Manual, Version 3.2, The Rand Corporation, Santa Monica

72 Jensen P (1983) Comp Phys Rep 1:1

CHAPTER II

NUCLEAR MOTION

§.5. Born-Oppenheimer Approximation.

The motion of a molecule may be thought of as collective motion due to its constituent particles. Often observables are explained based on the assumption that the motion of the electrons depends only parametrically on the nuclear positions.

Born and Oppenheimer [1] used perturbation theory in order to demonstrate that the separation of electronic and nuclear motion could take place if the expansion parameter κ, which equals $(m_e/m_N)^{1/4}$, was small. The particular choice of κ was motivated by the fact that the expansion of the full non-relativistic Schrödinger equation in terms of κ ensures the first and third order terms vanish. Hence by ignoring fourth and higher-order terms, the vibrational energy for small amplitudes about the equilibrium geometry is of second order in κ, whereas the rotational energy is of fourth order in κ.

The Born-Oppenheimer approximation leads to the well-known separated equations,

$$(\hat{T}_e + \hat{V}_e)\ \Psi_e\ (r_e, R_N) = E_e(R_N)\ \Psi_e\ (r_e, R_N)$$

$$(\hat{T}_N + \hat{E}_e(R_N) + \hat{V}_N)\Phi\ (R_N) = E_N(R_N)\ \Phi\ (R_N) \tag{5.1}$$

where r_e and R_N are the position vectors of all the electrons and nuclei respectively. Hence the electronic Schrödinger equation depends parametrically on the nuclear geometry.

Within this approach, the potential energy function for nuclear motion is described by the sum,

$$E(R_N) = E_e(R_N) + V_N(R_N) \tag{5.2}$$

This approximation is often referred to as the "clamped nuclei" approximation.

Embedded in the Born-Oppenheimer derivation is the assumption that nuclear motions have only a small amplitude around some nuclear configuration (usually assumed to be the equilibrium geometry). On the other hand, Born [2-3] has provided a much more general derivation, which is applicable to any geometry and moreover, makes no assumptions with respect to the amplitude of nuclear motion.

Born derived a non-relativistic Schrödinger equation in which he could decouple electronic from nuclear motion by setting the off-diagonal coupling terms to zero, obtaining the following nuclear equation,

$$\{\hat{T}_N + \hat{E}_e + \hat{B}_{\alpha\alpha}(R_N) + \hat{V}_N\}\Phi_{\alpha\beta}(R_N) = E\ \Phi_{\alpha\beta}(R_N) \tag{5.3}$$

where α indexes the different electronic states and β represents the nuclear solutions for each electronic state.

As Sutcliffe [4] has pointed out, the Hamiltonian as given by equation (5.3) still contains the continuous spectrum given by the centre-of-mass motion of the whole system. Furthermore, even if the translation motion is factored out of the eigenvalue equations, neither approach separates the remaining nuclear motion into vibration and rotation.

In equation (5.3) the diagonal electronic-nuclear coupling operator is given by,

$$\hat{B}_{\alpha\alpha}(R_N) = (1/2)\sum_{a=1}^{N} m_a^{-1} <\Phi_{\alpha\beta}|\Delta_a|\Phi_{\alpha\beta}> \tag{5.4}$$

Equation (5.3) is called the "adiabatic" approximation, since the potential energy still has a dependence not only on the instantaneous nuclear configuration, but also on the nuclear linear momenta for the particular electronic state under consideration. The non-adiabatic effects incorporate interactions with other electronic states which, for example, may be induced by vibrational and rotational motions of the nuclei [5]. Strictly speaking, it should be realised that equation (5.1) does not follow from a variation principle. However, Epstein [6] has shown that the exact lowest eigenenergy of equation (5.3) forms a lower bound to equation (5.1). Clearly the adiabatic scheme is only valid if the potential surface, as given by equation (5.3), is well separated from all other potential surfaces. The non-adiabatic coupling between states belonging to well separated surfaces has been largely ignored. However, care must be taken since it has been shown [7] that discrete states from a higher potential energy surface, as given by equation (5.3), may interact with continuum states belonging to another potential energy surface via a non-adiabatic coupling scheme, leading to the formation of an additional bound state. Hence the whole concept of a potential energy surface for nuclear motion has meaning only if non-adiabatic coupling terms can be ignored for the molecular state under consideration and moreover, for computational expediency if the diagonal terms are set to zero. It is important to re-emphasise that since there is an additional operator in the adiabatic eigenvalue equation, the two approaches (clamped nuclei and adiabatic) do not generally lead to the same solutions. Only if the diagonal coupling terms are also omitted are the two approaches identical.

The impact of the Born-Oppenheimer approximation in the case of angular triatomic molecules has been treated by Bunker and Moss [8] and Makushkin et al. [9]. Using perturbation theory, Makushkin et al. [9] showed that the leading adiabatic and non-adiabatic contributions to the vibrational and rotational spectral parameters are of order κ^6 and κ^8 with respect to the electronic energy and so are of the same order of magnitude as the sextic terms in the potential energy (the latter of which are only known for a handful of polyatomic molecules [10]). It is therefore difficult to separate the adiabatic and non-adiabatic effects from the higher-order contributions to the potential energy since they have similar orders of κ.

The development presented in this dissertation will assume that the clamped nuclei approximation holds (equation (5.2)) since there exists no direct experimental evidence that the adiabatic and non-adiabatic corrections are significant in vibration-rotation spectra of bent triatomic molecules in non-degenerate electronic ground states. To our knowledge, such an assumption has been generally borne out by experimental spectra of massive molecules. Nevertheless, it is important to be constantly mindful of this limitation [11], especially since the precision and accuracy of experimental data is rapidly improving.

§.6. Eckart-Watson Hamiltonian.

Over a number of decades spectroscopists viewed nuclear motion in terms of small amplitude vibrations and near rigid rotations about some vibrationally averaged or equilibrium geometry (the latter usually being well defined by a potential energy surface). In the early work, coordinate systems and nuclear Hamiltonians were selected using this framework. Hence the existence of an equilibrium geometry was embedded in the derivation of the kinetic energy operator. This implication has its roots in the formulation of a rovibrational Hamiltonian by Eckart [12], Wilson and Howard [13], Darling and Dennison [14] and Watson [15-16].

The quantum mechanical Hamiltonian is obtained by replacing the classical position and momenta (in cartesian coordinates) by positional and differential operators. That is, for a system with total energy given by,

$$E(q_\alpha) = \frac{1}{2m} \sum_\alpha p_\alpha^2 + V(q_\alpha) \tag{6.1}$$

The quantum mechanical Hamiltonian is obtained using the transformations,

$$p_\alpha \rightarrow -i\hbar \frac{\partial}{\partial q_\alpha} \tag{6.2}$$

$$q_\alpha \rightarrow \hat{q}_\alpha \tag{6.3}$$

to give,

$$\hat{H}(q_\alpha) = -\frac{\hbar^2}{2m} \sum_\alpha \frac{\partial^2}{\partial q_\alpha{}^2} + V(q_\alpha) \tag{6.4}$$

However, when this transformation is applied directly in coordinate systems other than the cartesian coordinate system, spurious results are obtained. Podolsky [17], Epstein [18] and Dirac [19] have shown how the correct quantum mechanical Hamiltonian can be obtained from a generalised coordinate system.

Podolsky [17] developed a Hamiltonian of a conservative system satisfying all the requirements of quantum mechanics for an arbitrary coordinate system. For a system where the kinetic energy is given by,

$$T = \frac{1}{2} \sum_{k=1}^{n} \sum_{l=1}^{n} g^{kl} p_k p_l \tag{6.5}$$

Podolsky's quantum mechanical Hamiltonian is,

$$\hat{H} = \frac{1}{2m} \sum_{r=1}^{n} \sum_{s=1}^{n} g^{-1/4} \hat{p}_r g^{1/2} g^{rs} \hat{p}_s g^{-1/4} + \hat{V} \tag{6.6}$$

where g^{-1} is the determinant of the matrix with elements g^{rs} and the $\hat{p}_r$ are the momentum operators given by,

$$\hat{p}_r = -i\hbar \frac{\partial}{\partial q_r} \tag{6.7}$$

The wavefunctions obtained using Podolsky's form of the nuclear Hamiltonian are normalised. That is,

$$\int \ldots \int \Psi_i^* \Psi_i \, dq_1 dq_2 dq_3 \ldots = 1 \tag{6.8}$$

where Ψ are the rovibrational wavefunctions.

Schaad and Hu [20] have confirmed that Podolsky's form of the quantum mechanical Hamiltonian is correct. In addition, they have discussed the relationship of other Hamiltonians, within generalised coordinate systems, to Podolsky's Hamiltonian.

Eckart developed the coordinate system for normal, non-linear molecules which is most suitable for solution of the rovibrational problem for small amplitude vibrations [12]. The system is based on molecule-fixed rotating coordinates and is designed to reduce the coupling terms in the expression for the kinetic energy. Sayvetz [21] generalised the theory so that it is applicable to molecules of all geometries. In this coordinate system, the first order terms of the total angular momentum are small. The Eckart framework is defined by Figure 2.1 and is given as follows:

(i) Consider a system whereby the origin of a molecule-fixed coordinate system, "O", is located by a vector $\mathbf{R}$ from a laboratory-fixed origin.

(ii) Each atom of the molecule is located with respect to the molecule-fixed coordinate system by the vector $\mathbf{r}_i$, the equilibrium position in the molecule-fixed coordinate system being defined by $\mathbf{r}_i^0$.

(iii) The rotating coordinate system has an angular velocity of ω.

This framework thereby implies the existence of a molecular equilibrium geometry.

The velocity of the ith particle in this frame of reference is given by,

$$\dot{\mathbf{R}} + \omega \times \mathbf{r}_i + \dot{\mathbf{r}}_i \tag{6.9}$$

The total kinetic energy for a molecule with n atoms is therefore given by,

$$2T = \dot{\mathbf{R}}^2 \sum_{i=1}^{n} m_i + \sum_{i=1}^{n} m_i (\omega \times \mathbf{r}_i) \cdot (\omega \times \mathbf{r}_i)$$

$$+ \sum_{i=1}^{n} m_i \dot{\mathbf{r}}_i^2 + 2\dot{\mathbf{R}} \cdot \sum_{i=1}^{n} m_i \dot{\mathbf{r}}_i$$

$$+ 2\dot{\mathbf{R}} \times \omega \cdot \sum_{i=1}^{n} m_i \mathbf{r}_i + 2\omega \cdot \sum_{i=1}^{n} m_i \mathbf{r}_i \times \dot{\mathbf{r}}_i \tag{6.10}$$

The first term is the kinetic energy due to the translation of the molecule-fixed coordinate system, the second term is the sum of the rotational kinetic energy of each atom and the third term is the sum of the vibrational kinetic energy of each atom. The final three terms represent the kinetic energy due to the coupling among the three types of motion: translation; vibration and rotation.

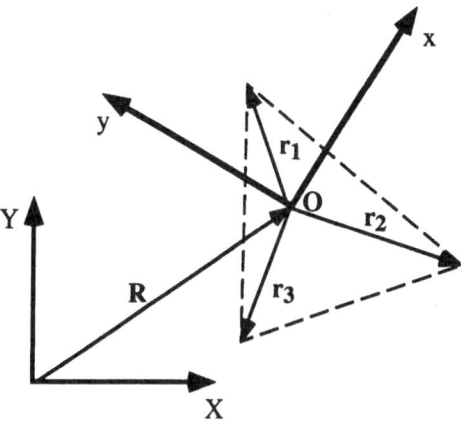

Figure 2.1 The Eckart framework. The origin, **O**, of the molecule-fixed x-y coordinate system, is located by a vector, **R**, from the origin of the laboratory-fixed X-Y coordinate system. The r_i vectors locate each atom with respect to the x-y coordinate system. The x-y coordinate system rotates with respect to the X-Y coordinate system with an angular velocity of ω.

The Sayvetz [21] rotating system is defined by imposing the conditions,

$$\sum_{i=1}^{n} m_i \, r_i = 0 \tag{6.11}$$

and

$$\sum_{i=1}^{n} m_i \, r_i^0 \times r_i = 0 \tag{6.12}$$

where m_i is the mass of the ith atom. The first of these conditions imposes that the centre-of-mass of the molecule remains at the origin of the moving coordinate system and the second condition ensures the removal of the first-order approximation of the angular momentum relative to the molecule-fixed system. Using,

$$\rho_i = r_i - r_i^0 \tag{6.13}$$

and the Sayvetz conditions, the kinetic energy becomes,

$$2T = R^2 \sum_{i=1}^{n} m_i + \sum_{i=1}^{n} m_i \, (\omega x \, r_i) \cdot (\omega x \, r_i) + \sum_{i=1}^{n} m_i \, \dot{\rho}_i^2$$

$$+ \, 2\omega \cdot \sum_{i=1}^{n} m_i r_i \times \dot{\rho}_i \tag{6.14}$$

By employing the Sayvetz conditions the coupling between the translational/rotational and between translational/vibrational motion are removed. Furthermore, the first-order coupling between vibrational and rotational motion is zeroed. Since the translational motion is no longer coupled, the translational kinetic energy term can be separated and is omitted hereafter. That is, the kinetic energy becomes,

$$2T = \sum_{i=1}^{n} m_i \, (\omega \times r_i) \cdot (\omega \times r_i) + \sum_{i=1}^{n} m_i \, \dot{\rho}_i^2$$

$$+ \, 2\omega \cdot \sum_{i=1}^{n} m_i \, r_i \times \dot{\rho}_i \tag{6.15}$$

This can be expressed in the form,

$$2T = \sum_{\alpha} \sum_{\beta} I_{\alpha\beta} \, \omega_\alpha \, \omega_\beta + \sum_{i=1}^{n} \sum_{\alpha} m_i \, \dot{\rho}_{\alpha i}^{\,2}$$

$$+ \, 2 \sum_{\alpha} \sum_{\beta} \sum_{\gamma} e_{\alpha\beta\gamma} \omega_\alpha \sum_{i=1}^{n} m_i \, r_{\beta i} \, \dot{\rho}_{\gamma i} \tag{6.16}$$

where Greek subscripts are used represent x, y, z directions of the moving coordinate system. Here $e_{\alpha\beta\gamma}$ is the permutation symbol and is only non-zero if α, β and γ are all different and is +1 if they are x, y, z in cyclic order and -1 if they are x, y, z in anti-cyclic order. The moment of inertia, $I_{\alpha\beta}$, is given by,

$$I_{\alpha\beta} = \sum_{\gamma} \sum_{\delta} \sum_{\varepsilon} e_{\alpha\gamma\varepsilon} \, e_{\beta\delta\varepsilon} \sum_{i=1}^{n} m_i \, r_{\gamma i} r_{\delta i} \tag{6.17}$$

Using the Eckart framework, Wilson and Howard [13] developed a quantum mechanical Hamiltonian which resembled that of Podolsky, but differed in the normalisation imposed on the wavefunctions. Following the work of Wilson and Howard [13], Darling and Dennison [14] rederived the quantum mechanical Hamiltonian using a generalised coordinate system. The

Darling and Dennison Hamiltonian is hermitian and the wavefunctions are normalised following Podolsky [17].

The motion of an n atom molecule in the Eckart frame can be described by 3n coordinates. In the Wilson and Howard [13] and the Darling and Dennison [14] Hamiltonian, three translational coordinates are used to represent the motion of the centre-of-mass of the molecule-fixed system with respect to the laboratory-fixed system, Euler angles are used to represent the rotation of the molecule and vibrational normal coordinates represent its vibration. The derivation of the quantum mechanical Hamiltonian involves initially expressing the displacements of the molecule within the molecule-fixed coordinate system in terms of N mass-weighted normal coordinates Q_i. For a non-linear molecule containing n atoms, N is equal to 3n-6, whereas for a linear molecule N is equal to 3n-5. That is,

$$\rho_{\alpha i} = m^{1/2} \sum_{m=1}^{N} c_{\alpha im} Q_m \tag{6.18}$$

where the normalisation condition implies,

$$\sum_{i=1}^{n} \sum_{\alpha} c_{\alpha im} c_{\alpha in} = \delta_{mn} \tag{6.19}$$

Transforming the displacement coordinates in the kinetic energy expression to normal coordinates gives,

$$2T = \sum_{\alpha} \sum_{\beta} I_{\alpha\beta}\, \omega_\alpha\, \omega_\beta + \sum_{m=1}^{N} \dot{Q}_m^2$$

$$+ 2 \sum_{\alpha} \sum_{\beta} \sum_{\gamma} e_{\alpha\beta\gamma} \omega_\alpha \sum_{i=1}^{n} \sum_{m=1}^{N} \sum_{n=1}^{N} c_{\beta im} c_{\gamma in}\, Q_m \dot{Q}_n \tag{6.20}$$

The Coriolis coupling term is defined as,

$$\zeta_{mn}^{\alpha} = \sum_{\beta} \sum_{\gamma} e_{\alpha\beta\gamma} \sum_{i=1}^{n} c_{\beta in} c_{\gamma in} \tag{6.21}$$

Substituting this into the expression for the kinetic energy gives,

$$2T = \sum_{\alpha} \sum_{\beta} I_{\alpha\beta} \, \omega_{\alpha} \, \omega_{\beta} + \sum_{m=1}^{N} Q_m^2$$

$$+ 2 \sum_{\alpha} \omega_{\alpha} \sum_{m=1}^{N} \sum_{n=1}^{N} \zeta_{mn}^{\alpha} Q_m \dot{Q}_n \qquad (6.22)$$

In order to obtain the quantum mechanical Hamiltonian the kinetic energy must be expressed in terms of momenta rather than velocities. The total angular momentum is given by,

$$\Pi_{\alpha} = \frac{\partial T}{\partial \omega_{\alpha}} = \sum_{\beta} I_{\alpha\beta} \omega_{\beta} + \sum_{m=1}^{N} \sum_{n=1}^{N} \zeta_{mn}^{\alpha} Q_m Q_n \qquad (6.23)$$

The momentum conjugate to the normal coordinates is given by,

$$P_m = \frac{\partial T}{\partial \dot{Q}_m} = \dot{Q}_m + \sum_{\alpha} \omega_{\alpha} \sum_{n=1}^{N} \zeta_{mn}^{\alpha} Q_n \qquad (6.24)$$

Substituting (6.24) into (6.23) gives,

$$\Pi_{\alpha} = \sum_{\beta} \omega_{\beta} I'_{\alpha\beta} + \pi_{\alpha} \qquad (6.25)$$

where $I'_{\alpha\beta}$ is the effective moment of inertia and π_{α} is the vibrational contribution to the total angular momentum.

If μ is defined as the inverse of I', then,

$$\omega_{\beta} = \sum_{\beta} \mu_{\alpha\beta} (\Pi_{\beta} - \pi_{\beta}) \qquad (6.26)$$

Since the determinant of I' is equal to zero for a linear molecule, the inverse does not exist and this expression is not correct. The formulation of the problem for linear molecules therefore deviates from that which follows. Various forms of the quantum mechanical Hamiltonian for a linear molecule have been obtained [16, 22]. However, we will restrict ourselves to non-linear molecules here.

The expression for the kinetic energy of a non-linear molecule becomes,

$$2T = \sum_{m=1}^{N} P_m^2 + \sum_{\alpha} \sum_{\beta} \omega_\alpha \omega_\beta I_{\alpha\beta}' \tag{6.27}$$

Using equation (6.26) to substitute for the angular velocity,

$$2T = \sum_{m=1}^{N} P_m^2 + \sum_{\alpha} \sum_{\beta} \mu_{\alpha\beta} (\Pi_\alpha - \pi_\alpha)(\Pi_\beta - \pi_\beta) \tag{6.28}$$

The kinetic energy is now in terms of momenta and can be used to obtain the quantum mechanical Hamiltonian. In matrix notation, the expression above can be written,

$$2T = \mathbf{PGP}^T \tag{6.29}$$

where,

$$\mathbf{G} = \begin{bmatrix} \mu_{xx} & \mu_{xy} & \mu_{xz} & 0 & 0 & 0 & \cdots \\ \mu_{yx} & \mu_{yy} & \mu_{yz} & 0 & 0 & 0 & \cdots \\ \mu_{zx} & \mu_{zy} & \mu_{zz} & 0 & 0 & 0 & \cdots \\ 0 & 0 & 0 & 1 & 0 & 0 & \cdots \\ 0 & 0 & 0 & 0 & 1 & 0 & \cdots \\ 0 & 0 & 0 & 0 & 0 & 1 & \cdots \\ \vdots & \vdots & \vdots & \vdots & \vdots & \vdots & \end{bmatrix} \tag{6.30}$$

and,

$$\mathbf{P} = [\Pi_x - \pi_x , \Pi_y - \pi_y , \Pi_z - \pi_z , P_1 , P_2, P_3, \dots] \tag{6.31}$$

Equation (6.29) is in the general form for the use of Podolsky's transformation [17]. The quantum mechanical Hamiltonian for the non-linear molecule within the Eckart frame is therefore given by,

$$\hat{H} = \frac{1}{2} \sum_{\alpha} \sum_{\beta} \mu^{1/4} (\hat{\Pi}_\alpha - \hat{\pi}_\alpha) \mu_{\alpha\beta} \mu^{-1/2} (\hat{\Pi}_\beta - \hat{\pi}_\beta) \mu^{1/4}$$

$$+ \frac{1}{2} \sum_{k=1}^{N} \mu^{1/4} \hat{P}_k \mu^{-1/2} \hat{P}_k \mu^{1/4} + \hat{V} \tag{6.32}$$

Watson simplified the form of this Hamiltonian [15] to give the simplest quantum-mechanical form of Eckart's classical rovibrational Hamiltonian [12]. By use of commutation relationships of operators within the Hamiltonian, Watson [15] proved that for non-linear molecules the order of factors in the first term is immaterial because of the relation,

$$\sum_{\alpha\beta} \pi_\alpha \mu_{\alpha\beta} = \sum_{\alpha\beta} \mu_{\alpha\beta} \pi_\alpha \tag{6.33}$$

Hence the rovibrational Hamiltonian can be compactly expressed as,

$$\hat{H} = \frac{1}{2} \sum_\alpha \sum_\beta (\hat{\Pi}_\alpha - \hat{\pi}_\alpha) \mu_{\alpha\beta} (\hat{\Pi}_\beta - \hat{\pi}_\beta) + \frac{1}{2} \sum_{k=1}^{N} \hat{P}_k^2$$

$$- \frac{1}{8} \hbar^2 \sum_\alpha \mu_{\alpha\alpha} + \hat{V} \tag{6.34}$$

In this expression, the third term gives a mass-dependent contribution to the potential energy and is referred to as the Watson operator.

Initially the rovibrational Schrödinger equation was solved using perturbation theory. In this treatment the accuracy of the solutions obtained is dependent on the expansion order being considered and the solution may no longer be an upper bound to the exact energy. Hence the variational approach is preferred.

§.7. Watson's Reduced Hamiltonians.

The expanded form of equation (6.34) is obtained by expanding $\mu_{\alpha\beta}$ and the potential energy V about the equilibrium geometry in terms of the normal coordinates q. The Hamiltonian is then expanded as,

$$\hat{H} = \sum_k \sum_l \hat{H}_{kl} \tag{7.1}$$

where the first and second subscripts label the order in vibrational operators and components of the total angular moment operator respectively. A simple physical meaning can be ascribed to the various operators in equation (7.1). For example, $\hat{H}_{02}$ is the rigid rotor approximation, $\hat{H}_{12}$ and $\hat{H}_{22}$ are the centrifugal distortion operators, $\hat{H}_{21}$ describes the Coriolis interaction between rotation and vibration, $\hat{H}_{20}$ is the Harmonic oscillator, $\hat{H}_{30}$ and $\hat{H}_{40}$ describe the anharmonicity of molecular vibrations. Watson [23] and Papoušek and Aliev [24] have detailed the various terms in the expansion of equation (7.1).

In the inverse secular problem, the strategy is to fit the observed rotational spectrum (i.e. eigenvalues of the rotational Hamiltonian) in order to determine the coupling parameters (rotational and centrifugal distortion constants) that occur in the rotational Hamiltonian itself. However, in this type of problem, the observables may depend on a certain combination of the parameters and so tend to be indeterminate. The task is to search for these indeterminancies and eliminate them, in order to obtain the same eigenvalues, but with a smaller number of parameters. That is, the reduced rotational Hamiltonian $\hat{H}_{ROT}^{RED}$ can be obtained by the contact transformation,

$$\hat{H}_{ROT}^{RED} = \hat{U}^{-1} \hat{H}_{ROT} \hat{U} \qquad (7.2)$$

where $\hat{U}$ is a unitary operator and $\hat{H}_{ROT}$ has as many parameters eliminated as possible in order to remove the indeterminancy.

The theory of quartic centrifugal distortions in asymmetric top molecules was initially developed by Kivelson and Wilson (KW) [25] and later put on more solid footing by Watson, who in a series of papers [26-28] showed that KW expression for the quartic distortion Hamiltonian contains an indeterminancy, since one term in $\hat{H}_{04}$ can be eliminated to give five determinable quartic coefficients.

Adopting a more convenient notation, the rotation part of the A-reduced asymmetric quartic Hamiltonian given by Watson [23] is,

$$\hat{H}_{ROT}^{RED} = A\hat{J}_a^2 + B\hat{J}_b^2 + C\hat{J}_c^2 - \Delta_J \hat{J}^4 - \Delta_{JK}\hat{J}^2\hat{J}_a^2 - \Delta_K \hat{J}_a^4 - 1/2[\delta_J\hat{J}^2 + \delta_k\hat{J}_a^2, \hat{J}_+^2 + \hat{J}_-^2]_+ + \ldots \qquad (7.3)$$

where the angular momentum operators Js are measure in units of $\hbar$ and refer to the principal axes of the rotating molecule. The axes are labelled a, b, c when the rotational constants are in order of decreasing magnitude such that $A \geq B \geq C$. The anti-commutator $[A, B]_+$ represents the relationship **AB+BA**.

The relationships between the unreduced Hamiltonian quartic terms and the A-reduced Hamiltonian centrifugal distortion terms are given by [23],

$$\tau_{AAAA} = -4 (\Delta_J + \Delta_{JK} + \Delta_k)$$

$$\tau_{BBBB} = -4 (\Delta_J + \delta_J) \qquad (7.4)$$

$$\tau_{CCCC} = -4 (\Delta_J - \delta_J)$$

For a planar molecule the relations for the quartic distortion constants are [24],

$$\tau_{CCCC} = (C_e/A_e)_4\tau_{AAAA} + (C_e/B_e)^4\tau_{BBBB} + 2(C_e^2/A_eB_e)^2\tau_{AABB}$$

$$\tau_{AACC} = (C_e/A_e)_2\tau_{AAAA} + (C_e/B_e)^2\tau_{AABB} \qquad (7.5)$$

$$\tau_{BBCC} = (C_e/A_e)_2\tau_{AABB} + (C_e/B_e)^2\tau_{BBBB}$$

For a molecule of C_{nv} symmetry with the x-axis in the σ_v plane the general form of $\hat{H}_{ROT}$ in the quartic approximation given by KW is [25],

$$\hat{H}_{ROT} = B_x\hat{J}^2 + (B_z-B_x)\hat{J}_z^2 - D_J(\hat{J}^2)^2 - D_{JK}\hat{J}^2\hat{J}_z^2 - D_K\hat{J}_z^4..... \qquad (7.6)$$

The relationship between equations (7.3) and (7.6) has been given by Watson [23] and are,

$$\Delta_J = D_J - 2R_6$$

$$\Delta_K = D_{JK} - 10\,R_6 \qquad (7.7)$$

$$\delta_K = -2R_5 - 4\sigma R_6$$

$$\Delta_{KJ} = D_{JK} + 12R_6$$

$$\delta_J = \delta_J$$

where σ is an asymmetry parameter given by,

$$\sigma = (2B_z-B_x-B_y)/(B_x-B_y) \qquad (7.8)$$

A useful and related asymmetry parameter has been given by Ray [29]. It is defined as,

$$\kappa = (2B-A-C)/(A-C) \qquad (7.9)$$

It is easy to see that κ must lie in the range,

$$-1 \leq \kappa \leq +1 \qquad (7.10)$$

and so κ equal to -1 corresponds to the limiting case for a prolate symmetric top (B=C), whereas if κ is equal to +1 the limiting case is that for an oblate top (A=B).

For K_2Li^+ spectroscopic constants were obtained using a least-squares fit to ab initio variationally determined rovibrational eigenenergies [30], the results of which are given in Table 2.1. The limiting case for the rotational levels of K_2Li^+ is Mulliken's prolate symmetric top. This is reflected by the Ray's asymmetry parameter, which is calculated to be -0.99 for the lowest lying five vibrational states. Hence the rotational energy levels were assigned within this framework and were calculated to the J equal 5 level [30].

REFERENCES TO CHAPTER II

1 Born M, Oppenheimer R (1927) Ann Physik 84:457

2 Born M, Huang K (1954) Dynamical theory of crystal lattices, Oxford University Press, London

3 Born M (1951) Gött Nachr Math Phys K1 1

4 Sutcliffe BT, Tennyson J (1987) J Chem Soc Faraday Trans 2 83:1663

5 Fernandez FM, Ogilvie JF (1992) Chin J Phys 30:177

6 Epstein ST (1966) J Chem Phys 44:4062

7 Rosenfield J, Voigt B, Mead CA (1970) J Chem Phys 53:1960

8 Bunker PR, Moss RE (1980) J Mol Spectrosc 80:217

9 Makushkin YS, Terent'ev AV, Ulenikov ON (1976) Nauka Novosibirsk

10 Dinelli BM, Crofton MW, Oka T (1988) J Mol Spectrosc 127:1

11 Teffo JL (1993) Mol Phys 78:1493

12 Eckart C (1935) Phys Rev 47:552

13 Wilson EB, Howard JB (1936) J Chem Phys 4:260

14 Darling BT, Dennison DM (1940) Phys Rev 57:128

15 Watson JKG (1968) Mol Phys 15:479

16 Watson JKG (1970) Mol Phys 19:465

17 Podolsky B (1928) Phys Rev 32:812

18 Epstein PS (1926) Proc Nat Acad Sci 12:634

19 Dirac PAM (1927) Proc Royal Soc A 113:621

20 Schaad LG, Hu J (1989) J Mol Struct (Theochem) 185:203

21 Sayvetz A (1939) J Chem Phys 7:383

22 Nielsen HH (1951) Rev Mod Phys 23:90

23 Watson JKG (1977) In: Vibrational spectra and structure, Durig JR (Ed) Vol 6, Elsevier, Amsterdam

24 Papoušek D, Aliev MR (1982) Molecular vibrational-rotational spectra, Elsevier, Amsterdam

25 Kivelson D, Wilson EB (1952) J Chem Phys 20:1575

26 Watson JKG (1967) J Chem Phys 46:1935

27 Watson JKG (1968) J Chem Phys 48:181

28 Watson JKG (1968) J Chem Phys 48:4517

29 Ray BS (1932) Z Physik 78:74

30 Wang F, Searles DJ, von Nagy-Felsobuki EI (1992) J Phys Chem 96:6158

Table 2.1 Spectroscopic Constants of K_2Li^+ for the Lowest Five Vibrational States (/MHz) [a].

v_{vib}	1	2	3	4	5
κ	-0.9898	-0.9892	-0.9887	-0.9885	-0.9874
A+C	0.1343+05	0.1335+05	0.1333+05	0.1320+05	0.1311+05
A-C	0.1177+05	0.1167+05	0.1163+05	0.1148+05	0.1139+05
τ_{AAAA}	-6.4404	-1.5958	0.2675	0.8000	0.8063
τ_{BBBB}	-0.0607	0.0279	-0.0812	0.2576	-0.3002
τ_{CCCC}	-0.0020	-0.1138	0.1363	-0.2823	0.2435
τ_{AACC}	5.0456	-14.8606	21.2536	-49.0669	46.7714
τ_{BBCC}	-0.0275	-0.0536	0.0442	-0.0526	0.0151
τ_{AABB}	-4.4635	14.9912	-21.2873	49.0631	-46.8315
Reduction Distortion Constants					
Δ_J	0.8126	0.1960	-0.0233	-0.1322	-0.0633
Δ_{JK}	-2.5766	-0.6072	0.0672	0.4106	0.2010
Δ_K	1.7644	0.4397	-0.0780	-0.2079	-0.1986
δ_J	0.3987	0.1050	-0.0218	-0.0339	-0.0692
δ_K	-1.2250	-0.2221	0.0229	0.1209	0.1835
First-order Centrifugal Distortion Constants					
D_J	0.8884	-0.7900	1.3130	-3.1656	2.8795
D_{JK}	-3.0314	5.3085	-7.9505	18.6111	-17.4557
D_K	2.1435	-4.4901	6.6034	-15.3749	14.5153
δ_J	0.3987	0.1015	-0.0218	-0.0339	-0.0692
R_5	0.6891	-0.8856	1.3401	-3.1289	2.8884
R_6	0.0379	-0.4930	0.6681	-1.5167	1.4714

a) Reproduced with permission from reference [30].

CHAPTER III

DISCRETE POTENTIAL ENERGY SURFACES

§.8. General Considerations.

A number of discrete potential energy surfaces are available in the literature for a variety of triatomic molecules [1-3]. The construction of a discrete potential energy surface is not a trivial exercise, since care must be taken in: the design of the basis set; the configuration interaction (CI) methodology employed; the design of the geometrical grid for the calculations. Of these, the latter is the possibly the least understood, although recently von Nagy-Felsobuki and coworkers [3-4] have generated grid points using an adaptive scheme based on the quadrature algorithm used for the potential energy integrator (see Chapters VI and VII).

Basis set design has been review by a number of authors over the past two decades. In particular, Dunning and Hay [5], Huzinaga [6], Poirier et al. [7] and Wilson [8] have addressed the problem of systematically constructing basis sets using gaussian type functions (GTFs) and/or problems associated with basis set superposition or completeness errors.

In general the quality of a basis set depends on several factors: the total number of contracted basis functions; the number of primitive functions that are used to assemble the contracted set; the choice of the exponents in the basis functions; the choice of the contracted coefficients; the different type of l-dependent functions (e.g. s, p, d, f...) that are included in the basis sets. Gaussian type functions (GTFs) are usually preferred over Slater type functions (STFs) because their properties simplify the evaluation of multi-centred coulomb and exchange integrals. A double zeta (DZ) basis set comprises of two basis functions per occupied atomic orbital, whereas a triple zeta (TZ) represents three basis functions per occupied atomic orbital. Polarisation functions are usually added to the basis set since they better define the valence electron distribution in an electric field. They have an l-type which is at least one l-type higher than what is present amongst valence basis sets. For example, hydrogen and carbon usually have p and d polarisation functions respectively added to the basis sets. Furthermore, the primitive and contracted basis sets are given in parenthesis, with a slash acting as the delimiter between the two.

Table 3.1 compares ab initio harmonic frequencies reported by Yamaguchi and Schaefer [9] with experimental [10-12] infrared harmonic frequencies for H_2O, CH_4 and H_2CO. They employed DZ, DZ+P and a [10s6p/4s4p] basis set for the heavy atom augmented by two polarisation functions, whereas for hydrogen a [6s/4s] basis set was used augmented by a single polarisation function (and denoted as EBS). The CI method used involved single and double

Table 3.1 Comparison of Ab Initio and Experimental Harmonic Frequencies for Small Molecules[a].

Symmetry	DZ/CISD	DZ+P/CISD	EBS/CISD	Exp.[b]
		H_2O		
a_1	3710	3967	3937	3834
a_1	1649	1693	1688	1647
b_1	3880	4088	4043	3943
		H_2CO		
a_1	3223	3074	-	2944
a_1	1878	1869	-	1764
a_1	1651	1596	-	1563
b_1	1324	1243	-	1191
b_2	3315	3155	-	3009
b_2	1349	1306	-	1287
		CH_4		
a_1	3001	3116	-	3026
e	1571	1601	-	1583
t	3129	3257	-	3157
t	1407	1397	-	1367

a) All entries are in cm^{-1}. See reference [9] for further details.

b) See references [10-12].

substitutions with certain restrictions imposed on the number of excitations for the more electron dense systems [9].

It is clear from the Table 3.1 that adding polarisation functions increases the frequencies in the case of water and methane but not for formaldehyde. Furthermore, the DZ/CISD values for water give the best agreement to experiment, even though the EBS is at a higher level of the theory. Yamaguchi and Schaefer [9] further noted that although the DZ+P/CISD errors tended to be larger than the DZ/CISD errors, the former level of the theory gave larger systematic errors. However, this fortuitous cancelling of errors may not lead to effective scaling parameters at any level of the theory. For example, at DZ/CISD level of theory for methane the fundamental vibrations are overestimated and underestimated, when compared with experiment. Generally at the HF level using an EBS yields harmonic vibrational frequencies too large by 10-15 per cent when compared with experiment [13].

The choice of the basis set and CI method is heavily dependent on the computer resources available. For example, a full CI expansion increases factorially with the number of electrons and orbitals, thereby necessitating the N-particle expansion to be truncated to a finite-order. The most commonly used truncation is to constrain the excitation level to single and double excitations (CISD) from either a single reference determinant or from a multi-configurational reference determinant (MC), the determinant being obtained from SCF solutions. Recent reviews of CI methodologies have been given by: Meyer et al. [14] on the self-consistent electron pair (SCEP) method; Ahlrichs and Scharf [15] on coupled electron pair approximation (CEPA); Peyerimhoff and Buenker [16] and Bruna and Peyerimhoff [17] on multi-reference double excitation-CI (MRD-CI); Pople and coworkers [18-20] on closed-shell single reference determinant single and double excitation CI (SD), couple clusters (CC) and other CI methodologies available in the GAUSSIAN package; Werner [21] and Shepard [22] on direct multi-configurational (MC) SCF and MC-CI methods; Roos [23] and Bauschlicher et al. [24] on the complete active space self-consistent field (CASSCF) method. It is beyond the scope of this dissertation to detail the every CI method used in the calculation of discrete ab initio potential energy surfaces. However, a brief resume will be given below only to highlight a number of salient features of a few schemes.

The standard CI approach consists of expanding the many electron wavefunction $\Psi(1,2,....n)$ in a linear combination of configuration-state functions (CSFs) Φ_k so that,

$$\Psi(1,2,....n) = \sum_{k=1}^{L} C_k \Phi_k(1,2,...n) \qquad (8.1)$$

where L is the number of CSFs included, Φ_k is a linear combination of totally antisymmetric determinantal states consisting of n independent orbital functions φ_i and Cs are the expansion coefficients. In multi-reference CI (MRCI) equation (8.1) is truncated and substitutions are made with respect to a specified set of configurations.

The number of distinct CSFs is given by Weyl's formula,

$$\text{Distinct CSFs} = \frac{2S+1}{n+1} \; {}^{n+1}C_{1/2N-S} \; {}^{n+1}C_{1/2N+S+1} \tag{8.2}$$

where N is the total number of electrons and S is the spin quantum number.

A full CI treatment with even supercomputer access may be prohibitive except for only the simplest molecules. This can be readily seen by the number of CSFs required for H_3^+ compared with the H_2O molecule using double zeta basis functions. For H_3^+ and H_2O the total numbers of distinct CSFs are 21 and 1,002,001 respectively. The numbers of distinct determinants are 36 and 4,008,004 respectively. The number of symmetry-adapted CSFs for a particular state is a smaller subset of the total number of distinct CSFs.

In the case of the Hyllerass-CI method (H-CI) the wavefunction is expanded to explicitly incorporate correlation via a power series expansion of the interelectronic distance. Hence, in the case of H_3^+ equation (8.1) collapses to,

$$\Psi_{H-CI} = \sum_{k=1}^{L} C_{0k}\Phi_k + r_{12} \sum_{k=1}^{L} C_{1k}\Phi_k \tag{8.3}$$

where r_{12} is the interelectronic distance and the subscripts of the expansion coefficients additionally denote the power of the r_{12} term. It has been shown that the incorporation of the second term in equation (8.3) is of outmost importance in correlation recovery. The wavefunction is no longer an expansion of orthonormal basis functions and so the generalised eigenvalue problem must be solved for the expansion coefficients. The equations necessary to calculate the various matrix elements use cartesian gaussian functions, which are used to solve the generalised eigenvalue problem for two-electron N centred molecules.

In order to make computations more tractable, equation (8.1) can be rearranged to yield an expression in terms of a single reference self-consistent field (SCF) determinant. In this case, the full CI (FCI) wavefunction has form,

$$\Psi_{CI} = \Phi_{SCF} + \sum_{ia} a_i^a \, \hat{t}_i^a \Phi_{SCF} + 1/4 \sum_{ijab} a_{ij}^{ab} \, \hat{t}_{ij}^{ab} \Phi_{SCF} + \ldots \tag{8.4}$$

where the SCF determinant is in terms of spin-orbitals with subscripts i, j, k... labelling those occupied and a, b, c... those unoccupied spin-orbitals. The $\hat{t}$ represent operators that promote electron(s) from occupied to unoccupied orbital(s) within the SCF determinant. The a_i^a... are arrays which involve the coefficients to be determined. Truncation of the wavefunction at the third term yields a CI wavefunction that only involves single and double substitutions. The appropriate energy functional is evaluated variationally and so yields an CISD energy which is an upper bound to the FCI energy. However, the CISD energy is not size-consistent, since the projection equations for say, a diatomic molecule at large separations, can be shown to have a linear dependence in the a arrays as well as to be equated to an expression which has a quadratic dependence on the same arrays [20]. Consequently, a size-correction formula such as that given by Davidson [25],

$$\Delta E = (1 - C_{SCF}^2)\Delta E_{SD} \qquad (8.5)$$

is often employed. The Davidson correction is an estimator for higher-order terms and therefore yields an energy which is no longer variational. Nevertheless, for H_3^+ (which has only two electrons) all the higher excitations in equation (8.1) are necessarily zero and so the CISD gives the correct FCI solutions.

The coupled clustered (CC) approach uses a substitution operator exponentially. The substitution operator can be expanded in a power series as,

$$e^{\hat{t}}\Phi = (1 + \hat{t} + \frac{1}{2}\hat{t} + \frac{1}{6}\hat{t} + ...)\Phi \qquad (8.6)$$

If $\hat{t}$ is truncated to a single and double substitution operator then CC(SD) would differ from equation (8.4) due to the non-linear terms. Hence CC(SD) includes higher-order correlation effects than the corresponding CI truncations.

Using a double zeta basis the electronic configuration of the water molecule at the HF level is: $1a_1^2 \, 2a_1^2 \, 1b_2^2 \, 3a_1^2 \, 1b_1^2 \, 4a_1^0 \, 2b_2^0 \, 2b_1^0 \, 5a_1^0 \, 3b_2^0 \, 6a_1^0 \, 4b_2^0 \, 7a_1^0 \, 8a_1^0$. For the $^1A_1$ ground electronic state of water, Saxe et al. [26] have shown that the number of CSFs for single excitations is 19, for doubles 341, for triples 2842, for quadruples 14475, for quintuples 41952, for sextuples 72365, for seven-fold excitations 71434, for eight-fold 40046, for nine-fold 11492 and finally for ten-fold 1506.

Table 3.2 gives an energy error analysis for the various CI levels with respect to H_2O [27-29]. Of all the CI methods given the MRCI-SD(7) calculation yields a surface which most closely parallels the FCI surface. Not surprisingly the error in the SCF calculation becomes progressively larger and at infinite separation is 495 mE_h. At this separation the error in the CISD level of theory is 77.8 mE_h, whereas for the MRCI-SD(7) and MRCI-SD(9) the error is 1.70 and

Table 3.2 Analysis of Various CI Levels for H_2O^a.

CI	No.of CSFs	R_e	$1.5R_e$	$2R_e$
(i) Single SCF determinant[b]				
SCF	1	148.03	210.99	310.07
CISD	361	7.85	22.3	60.4
Davidson's corr.		1.95	2.65	0.37
CISDT	3203	6.71	18.68	49.72
CISDTQ	17678	0.26	1.10	4.35
(ii) Many-Body Methods[c]				
CC(SD)		1.79	5.59	9.33
CC(SD)+T(4)		0.42	1.66	-2.75
(iii) MRCI-SD[d]				
7	7906	1.94	2.05	1.90
8	22644	0.38	0.49	0.54
9	52452	0.07	0.17	0.14
(iv) Full CI[b]				
FCI	256473	-76.157866	-76.014521	-75.905247

a) All entries are in milli-Hartrees and relative to the FCI energy, which is in Hartrees.

b) See reference [26, 27, 29].

c) See reference [28-29].

d) See reference [29].

0.13 mE$_h$ respectively. The Davidson correction is not variational and shows a non-constant behaviour as a function of R$_e$.

Hampel et al. [30] did a comparison of QCISD, CCSD and Brueckner CCD methods for N$_2$, CO, F$_2$ and HF using large basis sets. Of the CI algorithms investigated they found that the QCISD method yielded the smallest CPU durations. Moreover, they showed that the QCISD method calculated spectral parameters in best agreement with experiment, with R$_e$ deviating less than 0.001-0.004 Å and harmonic frequencies too large by only 30-60 cm^{-1}. Difluorine was exceptional, the error in R$_e$ being 0.02 Å and ω_e 100 cm^{-1} too large.

§.9. Case Study: The Discrete Potential Energy Surface of H$_3^+$.

The simplest triatomic species in chemistry is H$_3^+$. With three protons and two electrons it can be thought of as a protonated hydrogen molecule. It is considered to play an important role in interstellar chemistry as a proton donor via reactions such as,

$$H_3^+ + X \rightarrow HX^+ + H_2 \tag{9.1}$$

where X might be carbon monoxide or nitrogen. It has only recently been identified in interstellar space [31], although its rovibrational structure was identified by Oka and coworkers [32] in a laboratory in 1980. Recently, Carrington and coworkers [33] have observed some 27,000 transitions within 200 cm^{-1} from states near the dissociation limit in the predissociation infrared spectrum. Hence, the challenge remains for theoreticians to resolve these transitions using ab initio variational methods.

Well in advance of the laboratory detection of its rovibrational structure, Carney and Porter (CP) [34-35] modelled the shape and structure of H$_3^+$, its electronic, dipole and quadrupole surfaces as well as its vibrational spectrum. Later they were able to show [36] that the first few calculated lines in the P, Q and R branches of the vibration-rotation spectrum were within 1% of the observed values. However, their calculated v_2 value was 2516.08 cm^{-1}, some 5.3 cm^{-1} less than the experimental value. This lead to the development of a series of discrete potential surfaces aimed at resolving this discrepancy and designed to produce reliable predictions of transitions involving more highly excited states in the bending mode. However, it should not be forgotten that Oka [37] initially used Carney and Porter's spectroscopic constants in order to unravel the vibration-rotation spectra of H$_3^+$.

A number of different discrete surfaces have been constructed for H$_3^+$ [34-35, 38-43]. Since H$_3^+$ is a two electron system, the most tractable CI methodology is the direct CI method. Hence the surfaces discussed below are mostly in this vain.

CP [34-35] constructed a 74 point full CI surface using 21 floating gaussian type functions (FGTFs), with 15 in the plane and another six out of the plane FGTFs. Their CI grid contained 16 equilateral, 25 isosceles (non-equilateral) and 33 asymmetric geometries. The first few calculated lines in the P, Q, R branches of vibration-rotation spectra were within 1% of experiment. CP also employed a full augmented CI calculation (ACI) using a basis set of 33 FGTFs, with 27 in the plane and six out of the plane functions. The non-linear FGTFs parameters were roughly optimised [34-35]. The resulting ACI surface [2-3, 44] yielded vibrational energies consistently larger than the CI surface by ~1%.

Schinke, Dupuis and Lester (SDL) [39] constructed a 650 point full CI potential energy surface using [5s,3p] GTFs. The grid was systematically chosen using the scattering coordinates (R, r, θ), where R is the distance of the proton to the mid-point of the H-H bond, r is the H-H bond length and θ is the angle subtended by **R** and **r**. The ranges used for each geometric parameter were: $0 < R < 10a_0$; $0.6a_0 < r < 2.6a_0$; $0^0 < \theta < 90^0$. Of the 720 points generated only 650 energy calculations were performed. Tennyson and Sutcliffe [45], using this surface, obtained similar results to CP for the stretching modes, but for vibrations involving the E mode their results were lower by ~1% .

Dykstra and Swope (DS) [38] employed a [8s, 3p, 1d/ 6s, 3p, 1d] contracted GTFs (CGTFs) hydrogen basis in their SCEP calculation. Their primary basis set consisted of 63 GTFs, with 21 centred on each nucleus. The core s functions were taken from Huzinaga [46] and the p and d functions have exponents of magnitude 1.6, 0.5, 0.15 and 1.33 respectively. They designed a 69 point grid, the points of which were chosen to include energies equal to several vibrational quanta above equilibrium. Carney et al. [47] employed a power series fit to DS surface and found that the agreement with experiment is within 1 cm^{-1} for the fundamental vibrations of H_3^+. However, to achieve this agreement Carney et al. [47] found it necessary to omit points 56, 63, 65 and 66 of the discrete surface.

Burton, von Nagy-Felsobuki, Doherty and Hamilton (BFDH) [40] constructed a 78 point pseudo natural orbitals (PNO) CI [48] surface using 81 CGTFs. The grid parallel that of CP [35-36] with additional points selected either off axis or well away from the minimum. Their surface contains 18 equilateral, 26 isosceles and 34 asymmetric nuclear configurations, representing 112 distinct points on the energy hypersurface. They used a [5s,3p](1s,1p)(1d) basis set derived from a root basis, which Burton and coworkers [49-50] systematically constructed for H_2. For each hydrogen the 5s basis represents a 2, 1, 1, 1, 1 contraction of Huzinaga's [46] 6s hydrogen basis. The (1s,1p) functions are bond functions fixed at 0.59 a_0 from each nuclear centre and positioned along the direction of each bond. For contracted nuclear configurations, overlap was avoided by constraining each sp set to be at least 0.1 a_0 apart. The 1d functions had exponents derived from the work of Schaefer and Meyer [51] and are also bond functions, but are positioned at the mid-point of each bond.

The use and position of bond functions in the BFDH calculations arose from the analysis of basis set incompleteness [52]. If energy lowering was used as a criterion for incompleteness, then there were regions where the nuclear centred basis functions demonstrated insufficient flexibility. Hence, the BFDH basis was specifically designed to avoid degradation in basis set completeness with bond extension and compression so that an even-handed representation could be obtained for the whole bowl of the H_3^+ surface near the minimum.

For H_2, the BFDH basis set yields an SCF energy of -1.133404 E_h and a PNO-CI energy of -1.172483 E_h at the equilibrium bond length [40]. This accounts for ~96 % of the correlation energy as given by the exact energy of -1.174470 E_h obtained by Kolos and Wolniewicz (KW) [53]. It is estimated [40] that about half the absolute error of 440 cm^{-1} is due to the use of approximate pseudo natural orbitals.

In order to assess the sensitivity of the energy to variations in basis sets, BFDH [40] carried out additional PNO-CI calculations at breathe mode displacements of -0.57, 0.0 and 2.29 a_0 using: a) [8s,4p,1d/6s,4p,1d] (1s,1p)(1d) - a 108 component basis set; b) BFDH [6s,3p/5s,3p] (1s,1p) (1d) - a 81 component basis set; c) DS [8s,3p,1d/6s,3p,1d] - a 63 component basis set. Comparison with the (63, 81) component basis, showed that the 108 component basis yields a lower H_3^+ energy by (55, 241), (46,184) and (298, 55) cm^{-1} at those respective displacements. BFDH [40] concluded that while the DS basis yielded a lower CI energy near the minimum, the BFDH basis possibly gives a better description at large positive displacements, the latter being more important to vibration since the potential is a slow varying function in this direction from R_e.

The initial power series fits [40] using the BFDH surface yielded vibrational energies that deviated by as much as 10-30 cm^{-1} from those of CP [34-35] and from experiment [31-32, 37, 54]. However, a more recent fit of the BFDH surface by Martire and Burton (MB) [55] yielded an E mode fundamental frequency of 2521.6 cm^{-1}, which compares favourably with the experimental value of 2521.308(9) cm^{-1}[37, 54].

More recently, Meyer, Botschwina and Burton (MBB) [41] generated a 69 point potential energy surface. The grid contains energy points up to 25,000 cm^{-1} above the minimum. They employed a 10s, 4p, 2d CGTFs hydrogen basis in their full CI calculations. They used Huzinaga's 10s set [46], contracting the four inner most functions. The 4p and 2d exponents were roughly optimised with the exponent having magnitudes of 2.6, 0.8, 0.3, 0.12, 2.0 and 0.7 respectively. MBB [41] have compared their own H_2 curve in the region of 0.8 a_0 < R < 3.0 a_0 with the accurate curves of KW [53] and with full CI calculations using the basis sets of DS [38] and SDL [39]. They have concluded that the DS and SDL curves show absolute errors of 240-350 cm^{-1} and 500-850 cm^{-1} respectively, whereas their own curves are in error by only 140-200 cm^{-1} in this range. Moreover, using their spd basis set their curve produces almost a constant error for R > 1.2 a$_o$, whereas with the addition of a f function (exponent fixed at 2.0) the absolute errors are reduce to 100-150 cm^{-1}, but with the undesirable feature that the error varies rather strongly as a function of R. Hence, no f functions were included in their 69 point potential energy surface. MBB [41] also investigated augmenting their spd nuclear centred basis with molecule-centred 2s, 1p, 1d and 1f functions. While they found that the total energy was lowered by 68 cm^{-1} due to numerical difficulties which would be encountered for linear geometries, the auxiliary functions were not included in the construction of their H$_3^+$ potential energy surface.

MBB [41] compared their H$_2$ vibrational frequency with that calculated using the basis sets of KW [53] and DS [38]. The MBB and DS vibrational frequency is 0.7 and 7 cm^{-1} larger than that of the KW calculation, respectively (the latter yielded 4162.4 cm^{-1}). The experimental value is 4161.1 cm^{-1} [56]. With respect to H$_3^+$ the mean deviation of the lowest-lying seven fundamental transitions is less than 2 cm^{-1}, with errors for other predicted frequencies expected to be below 0.1%.

Frye, Preiskom, Lie and Clementi (FPLC) [42] have calculated a 69 point potential energy surface for H$_3^+$ using Hylleraas-type CI calculations. The grid was selected to be the same as DS [38]. For the full surface they used a [13s,3p,1d] set of uncontracted GTFs on each site, yielding a total of 84 basis functions. The potential energy surface is found to be substantially lower in absolute energy than all the previous ab initio calculations. They estimate that the minimum energy is only 8 cm^{-1} higher than the best theoretical value, moreover it is 150 cm^{-1} lower than the MBB potential. The weighted root-mean-square (rms) deviation of the fit by Lie and Frye [43] is only 0.16 cm^{-1}, which is substantially better than the a weighted rms of 0.6 cm^{-1} obtained by MBB [41]. Not surprisingly for the low-lying vibrational band origins of the isotopomers of H$_3^+$ the maximum deviation between the two surfaces is 8 cm^{-1} for the sixth vibrational band origin of H$_3^+$. Deviations for more massive isotopomers are substantially smaller. Moreover, on comparison with available experimental results the maximum deviation of the surface is only 4.2 cm^{-1} for the ν_2 mode of H$_3^+$ [43].

At the CI level, BFDH H$_2$ results give energy improvements of 1247 and 209 cm^{-1} compared with the results reported by Csizmadia et al. [57] and SDL [39] respectively. However, the DS, MBB and FPLC basis yield lower H$_2$ energies. Table 3.3 compares the ab initio calculations of H$_3^+$ from Hirschfelder's [58] pioneering work in 1938 through to FPLC H-CI calculations in 1990 [42]. All the calculations [34-35, 38-42, 57-78] determine that H$_3^+$ is of triangular structure and in fact, is of D$_{3h}$ symmetry. After 1965, most calculations yield a R$_{H-H}$ bond length between 1.64 and

Table 3.3 Comparison of Ab Initio Calculations of H_3^+.

Reference	Method (Basis Set)	E_{min} (/E_h)	R_{H-H} (/a_o)	ω_1(/cm^{-1})	ω_2/ω_3(/cm^{-1})
[58]	MO (3 STF)	-1.293	1.79	1550	1100
[59]	MO-CI (16 FSGTF)	-1.3185	1.66	3610	-
[60]	MO-CI (12 STF)	-1.3326	1.6575	3400	2850
[34-35]	SD-CI (21 FGTF)[a]	-1.33519	1.6585	3185	2516
[61]	MO-CI (42 GTF)	-1.3359	1.6600	-	-
[62]	MO-CI (3 GTF)	-1.33764	1.6504	3301	-
[63]	PNO-CI (36 CGTF)	-1.33793	1.6600	-	-
[64]	MO-CI (118 STF)	-1.3392	1.6390	3450	2850
[34-35,44]	ACI (33 FSGTF)[a]	-1.33936	1.6500	3220	2546
[57]	MO-CI (33 GTF)	-1.3397	1.6600	-	-
[65]	SD-CI (33 GTF)	-1.34022	1.6830	-	-
[39]	MO-CI (42 CGTF)[a]	-1.34023	1.6600	-	-
[66-67]	CI (42 GTF)	-1.34050	1.6406	-	-
[68]	CI (42 GTF)	-1.34050	1.6500	-	-
[40]	PNO-CI (81 CGTF)	-1.34188	1.6525	3189	2509
[69]	SCC (12 FGTF)	-1.34203	1.6504	-	-
[40]	PNO-CI (63 CGTF)[a]	-1.34250	1.65041	-	-
[40]	PNO-CI (108 CGTF)	-1.34272	1.65041	-	-
[38]	SCEP (63 CGTF)[a]	-1.34278	1.6504	-	-
[41]	SD-CI (87 CGTF)[a]	-1.34309	1.6501	3178	2519
[70]	CI (18 SP)[b]	-1.34335	1.6500	3272	2735
[41]	SD-CI (104 CGTF)	-1.34340	1.6504	-	-
[71]	SCC (24 CGLO)[c]	-1.343422	1.6504	-	-
[72]	H-CI (48 CGTF)	-1.343500	1.6504	-	-
[74]	Random Walk	-1.34376	1.6500	-	-
[42]	H-CI (84 CGTF)[a]	-1.34379	1.6499	3183	2521
[42]	H-CI (138 GTFs)	-1.343828	1.6500	-	-
[73]	Random Walk	-1.3439	1.6500	-	-

a) Discrete potential energy surface.

b) Singer polynomial.

c) Contracted gaussian lobe functions.

1.66 a_0 [34-35, 38-42, 57, 59-74]. The more sophisticated potential energy surfaces of the 80s and 90s yield a R_{H-H} bond length of 1.650 a_0 [38-42]. Clearly the FPLC H-CI calculation of [42-43] yields the lowest variational energy reported in the literature to date, being only 8 cm^{-1} above the expected minimum [42]. The random walk method (which possibly yields a H_3^+ energy very near the "true" electronic energy) is not readily amenable for rovibrational calculations and moreover, is not variational [74].

Table 3.3 also compares (where possible) the calculated fundamental vibrational band origins using the various discrete CI surfaces. However, it should be noted that these results are not independent of the analytical function employed to represent the ab initio surface (see Chapter IV). For example, MB [55] using the BFDH [40] discrete ab initio potential determined the fundamental asymmetric stretch to be 2521.6 (within 0.3 cm^{-1} of experiment [54]) which differs from a Simon-Parr-Finlan power series fit which gives a value of 2509 cm^{-1} [40].

Figure 3.1 compares the H_3^+ breathe mode potential energy cut of DS [38], BFDH [40], MBB [41] and FPLC [42] surfaces. The fitting was done employing a seventh-order polynomial on each curve relative to its own equilibrium geometry. From Figure 3.1 it is clear that all four surfaces are nearly parallel near the minimum, with the BFDH surface being the poorest on an energy criterion. The MBB and FPLC surfaces are more parallel at large displacements (e.g. ±0.30 a_0) although agreement is not uniform over entire domain of the breathe mode coordinate.

A further appraisal of the similarity of the BFDH and MBB discrete potential energy surfaces at small displacements can be made by comparing the experimental fundamental vibration frequencies obtained using similar potential energy functions and an identical variational solution algorithm. Table 3.4 shows the fundamental vibration frequencies for H_3^+ and D_2H^+ using Morse type expansions of the BFDH [40] and MBB [41] discrete surfaces. The variation solution algorithm is that of Tennyson and Sutcliffe [45]. The difference between the two surfaces is within 4 cm^{-1}, with the MBB surface yielding a more accurate result when compared with experiment. Table 3.4 also gives the rotational constants [54, 75] obtained from the Eckart Hamiltonian using perturbation theory [76] with those calculated by using the same variational method [45, 77-78]. The J≤4 rotational levels were fitted to standard Watson rovibrational Hamiltonians (see Chapter II). The agreement with experiment is extremely good for H_3^+ and D_2H^+ using either the MBB or the BFDH discrete surfaces. However, at larger displacements of 2.0 and -0.8 a_0 the BFDH surface is extrapolated to be considerably lower than the MBB surface by ~2500 and 35000 cm^{-1} respectively. Hence agreement for the highly excited states should not be as good. Miller and Tennyson [78] have shown that by employing the MBB potential they can determine rotational constants of the H_3^+ isotopomers (not only for the ground state but also for excited states) that are competitive with experiment and which can only be represented by resorting to sixth-order perturbation theory.

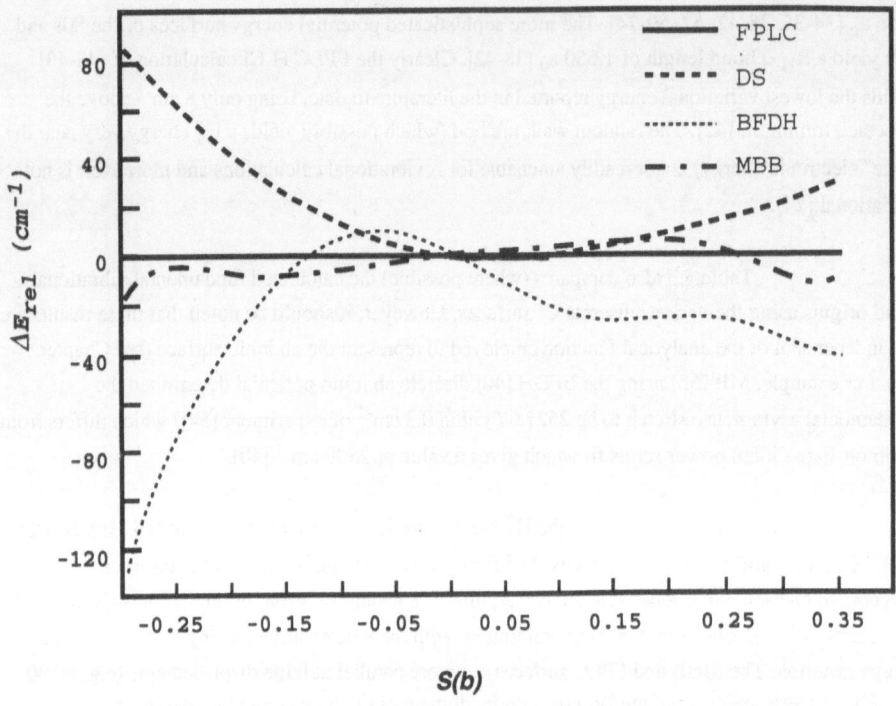

Figure 3.1 Difference between the H_3^+ ab initio breathe mode potential energy calculated by Frye et al. [42] and earlier ab initio calculations. The ordinate is displacements along the breathe mode. The curves presented are FPLC (——), DS (– – – –) [38], BFDH ($\cdots\cdots$) [40] and MBB (—–—) [41]. Reproduced with permission from reference [42].

Table 3.4 Comparison of Experimental and Calculated Vibration and Rotation Constants for the Ground State of H_3^+ and Isotopomers (in cm^{-1}).

Parameter	Exp[a]	MBB/TS[b]	BFDH/BM/TS[c]
		H_3^+	
ν_1	3178	3178	3175
ν_2/ν_3	2521	2521	2518
B	43.5646±0.0016	43.550	43.468
C	20.6051±0.0030	20.604	20.564
$10^2 D_{JJ}$	4.18±0.02	4.19	4.2
$10^2 D_{JK}$	-7.61±0.04	-7.71	-7.7
$10^2 D_{KK}$	3.74±0.04	3.84	3.8
$10^5 \delta_3$	-1.1±0.1	-0.7	-0.7
$10^3 d_J$	1.7±0.2	1.9	2.0
$10^3 d_K$	-4.1±0.1	-5.2	-5.3
		D_2H^+	
ν_1	2737	2737	2733
ν_2	1968	1968	1965
ν_3	2078	2079	2075
A	36.194±0.018	36.194	36.122
B	21.9039±0.0064	21.859	21.803
C	13.0856±0.0047	13.067	13.042
$10^2 D_{JJ}$	1.18±0.04	0.59	0.5
$10^2 D_{JK}$	-0.8±0.2	0.17	0.5
$10^2 D_{KK}$	[1.74][d]	1.98	1.7
$10^3 d_J$	0.225±0.06	0.21	1.0
$10^3 d_K$	1.6±0.3	1.22	5.6
$10^4 d_{JJJ}$	2.74±0.08	1.0	-

a) See reference [54, 75, 79].

b) See reference [55, 78].

c) See reference [40, 45, 55, 77].

d) Held fixed.

Of all the discrete ab initio surfaces mentioned, the CP [34-35], BFDH [40] , MBB [41] and FPLC [42] surfaces have been the most successful to date in unravelling the infrared spectrum of H_3^+. Of these four, the MBB surface [41] has been the most accurate at this time. However, it still remains to be seen whether the under utilised, but variationally lowest energy surface constructed by FPLC [42] will be shown to be of superior spectroscopic quality.

REFERENCES TO CHAPTER III

1 Dykstra CE (1988) Ab initio calculation of the structures and properties of molecules, Elsevier, Amsterdam

2 Tennyson J (1992) J Chem Soc Faraday Trans 2 88:3271

3 Searles DJ, von Nagy-Felsobuki, EI (1991) In: Vibrational spectra and structure, Durig JR (Ed), Vol 19, Elsevier, Amsterdam

4 Wang F, Searles DJ, von Nagy-Felsobuki EI (1992) J Phys Chem 96:6158

5 Dunning TH, Hay PJ (1977) In: Methods of electronic structure theory, Schaefer HF (Ed) Plenum, New York

6 Huzinaga S (1985) Comput Phys Rep 2:279

7 Poirier R, Kari R and Csizmadia IG (1985) Handbook of gaussian basis sets: a compendium for ab initio molecular orbital calculations, Elsevier, Amsterdam

8 Wilson S (1987) Adv Chem Phys 67:439

9 Yamaguchi Y, Schaefer HF (1980) J Chem Phys 73:2310

10 Bucknell MG, Handy NC (1974) Mol Phys 28:777

11 Duncan JL, Mallinson PD (1973) Chem Phys Lett 23:597

12 Gray DL, Robiette AG (1979) Mol Phys 37:1901

13 Pople JA, Schlegel HB, Krishnan R, DeFrees DJ, Binkley JS, Frisch MJ, Whiteside RA, Hout RF, Hehre WJ (1981) Int J Quant Chem Symp 15:269

14 Meyer W, Ahlrichs R, Dykstra CE (1984) In: Advanced theories and computational approaches to the electronic structure of molecules, Dykstra CE (Ed), Reidel, New York

15 Ahlrichs R, Scharf P (1987) Adv Chem Phys 67:501

16 Peyerimhoff SD, Buenker RJ (1980) In: Molecular physics and quantum chemistry into the 80s, Burton PG (Ed), University of Wollogong, Wollongong

17 Bruna PJ, Peyerimhoff SD (1987) Adv Chem Phys 67:1

18 Pople JA, Krishnan R, Schlegel HB, Binkley JS (1979) Int J Quant Chem Symp 13:225

19 Pople JA, Seeger R, Krishnan R (1977) Int J Quantum Chem Symp 11:149

20 Hehre WJ, Radom L, Schleyer PvR, Pople JA (1986) Ab initio molecular orbital theory, Wiley, New York

21 Werner HJ (1987) Adv Chem Phys 69:1

22 Shepard R (1987) Adv Chem Phys 69:63

23 Roos BO (1987) Adv Chem Phys 69:399

24 Bauschlicher CW, Langhoff SR, Taylor PR (1989) In: Supercomputer alogrithms for reactivity, dynamics and kinetics of small molecules, Lagana A (Ed), Kluwer Academic Publishers, Boston

25 Langhoff SR, Davidson ER (1974) Int J Quant Chem 8:61

26 Saxe P, Schaefer HF, Handy NC (1981) Chem Phys Lett 79:202

27 Harrison RJ, Handy NC (1983) Chem Phys Lett 95:386

28 Bartlett RJ, Sekino H, Purvis GD (1983) Chem Phys Lett 98:66

29 Brown FB, Shavitt I, Shepard R (1984) Chem Phys Lett 105:363

30 Hampel C, Peterson KA, Werner H-J (1992) Chem Phys Lett 190:1

31 Oka T (1980) Phys Rev Lett 45:531

32 Oka T, Geballe TR (1990) Astrophys J 351:L53

33 Carrington A, McNab IR (1989) Acc Chem Res 22:218

34 Carney GD, Porter RN (1974) J Chem Phys 60:4251

35 Carney GD, Porter RN (1976) J Chem Phys 65:3547

36 Carney GD, Porter RN (1980) Phys Rev Lett 45:537

37 Oka T (1983) In: Molecular ions:spectroscopy, structure and chemistry, Miller TA, Bondybey VE (Eds), North-Holland, Amsterdam

38 Dykstra CE, Swope WC (1979) J Chem Phys 70:1

39 Schinke R, Dupuis M, Lester WA (1980) J Chem Phys 72:3909

40 Burton PG, von Nagy-Felsobuki EI, Doherty G, Hamilton M (1985) Mol Phys 55:527

41 Meyer W, Botschwina P, Burton PG (1986) J Chem Phys 84:891

42 Preiskorn A, Lie GC, Frye D, Clementi E (1990) J Chem Phys 92:4948

43 Lie GC, Frye D (1992) J Chem Phys 96:6784

44 Carney GD (1980) Mol Phys 39:923

45 Tennyson J, Sutcliffe BT (1984) Mol Phys 51:887

46 Huzinaga S (1965) J Chem Phys 42:1293

47 Carney GD, Adler-Golden SM, Lesseski DC (1986) J Chem Phys 84:3921

48 Ahlrichs R, Driessler F, Lischka H, Staemmler V, Kutzelnigg W (1975) J Chem Phys 62:1235

49 Burton PG, Senff UE (1982) J Chem Phys 76:6073

50 Burton PG, Gray PD, Senff UE (1982) Mol Phys 47:785

51 Schaefer J, Meyer W (1979) J Chem Phys 70:344

52 Burton PG (1980) In: Molecular physics and quantum chemistry into the 80s, Burton PG (Ed), University of Wollongong, Wollongong

53 Kolos W, Wolniewicz L (1965) J Chem Phys 43:2429

54 Watson, JKG, Foster SC, McKellar ARW, Bernath P, Amano T, Pan SF, Crofton MW, Altman RS, Oka T (1984) Can J Phys 62:1875

55 Martire B, Burton PG (1985) Chem Phys Lett 121:479

56 Wolniewicz L (1966) J Chem Phys 45:515

57 Csizmadia IG, Kari RE, Polanyi JC, Roach AC, Robb MA (1970) J Chem Phys 52:6205

58 Hirschfelder JO (1938) J Chem Phys 6:795

59 Pearson AG, Poshusta RD, Browne JC (1966) J Chem Phys 44:1815

60 Christoffersen RE (1964) J Chem Phys 41:960

61 Kutzelnigg W, Ahlrichs R, Labib-Iskander I, Bingel WA (1967) Chem Phys Lett 1: 447

62 Schwartz ME, Schaad LJ (1967) J Chem Phys 47:5325

63 Ahlrichs R (1975) Theor Chim Acta 39:149

64 Borkman RF (1970) J Chem Phys 53:3153

65 Wright LR, Borkman RF (1982) J Chem Phys 77:1938

66 Duben AJ, Lowe JP (1971) J Chem Phys 55:4270

67 Duben AJ, Lowe JP (1971) J Chem Phys 55:4276

68 Kawaoka K, Borkman RF (1971) J Chem Phys 55:4637

69 Preiskorn A, Woźnicki W (1982) Chem Phys Lett 86:369

70 Salmon L, Poshusta RD (1973) J Chem Phys 59:3497

71 Preiskorn A, Woźnicki W (1984) Mol Phys 52:1291

72 Urdaneta C, Largo-Cabrerizo A, Lievin J, Lie GC, Clementi E (1988) J Chem Phys 88:2091

73 Mentch F, Anderson JB (1981) J Chem Phys 74:6307

74 Anderson JB (1987) J Chem Phys 86:2839

75 Lubic KG, Amano T (1984) Can J Phys 62:1886

76 Aliev M, Watson JKG (1985) In: Molecular spectroscopy, modern research III, Rao K (Ed) Academic Press, London

77 Sutcliffe BT, Tennyson J (1988) In: Maruani J (ed) Molecules in physics, chemistry and biology, Vol 2, Kluwer Academic Publishers, Boston

78 Miller S, Tennyson J (1987) J Mol Spectrosc 126:183

79 Majewski WA, Marshall MD, McKellar ARW, Johns JWC, Watson JKG (1987) J Mol Spectrosc 122:341

CHAPTER IV

POTENTIAL ENERGY FUNCTIONS

§.10. Historical Approach.

Calculations of spectroscopic parameters require knowledge of a functional representation of the electronic surface. Despite the problems associated with singularities [1] workers will continue to represent surfaces by functional forms. Consequently, there have been numerous functional representations proposed in the literature, with the review of Searles and von Nagy-Felsobuki [2] and monographs of Murrell and coworkers [3], Mezey [4] and Varandas [5] documenting some of these approaches.

In the case of a diatomic molecule the extraction of a potential surface from spectroscopic data is common place. Despite the number of embedded approximations, the Rydberg-Klein-Rees (RKR) method [3] (together with higher order corrections [6] or correction due to inverted perturbations [7]) yields remarkably reliable quantal potentials. However, recent improvements in experimental precision have tested even the best potentials, with the result that in order to realign theory with experiment a combination of functional forms rather than a single global representation may be required. That is, a functional form spliced from an analytical potential for the repulsive region, a RKR potential for the minimum/intermediate zone and a van der Waals potential for the asymptotic region [8].

Potentials from either ab initio electronic or RKR calculations are necessarily discrete and so provide the impetus to discover the underlying "universal" representation - if not globally then at least near the minimum energy of the surface. The search for the "Holy Grail of Spectroscopy" started with Dunham [9-10] in 1932 and has continued since that time.

The historical approach was to develop a power series representation involving the Dunham expansion variable, of form $(R-R_e)/R_e$ where R and R_e are the instantaneous and equilibrium internuclear distances. More recently, alternate potential functions have been developed. In the case of the Simon-Parr-Finlan (SPF) [11], Ogilvie [12], Thakkar [13] and Huffaker [14] potential functions, the expansions are in the powers of $(R-R_e)/R$, $2(R-R_e)/(R+R_e)$, $\{1-(R_e/R)^{-a-1}\}$ and $\{1-\exp(-a(R-R_e)\}$ respectively (where "a" is a Dunham constant). The latter expansion represents a perturbation of the well known Morse potential [15] and is based on the work of Coolidge, James, and Vernon [16]. Power series expansions are generally still used for global fits. They are used in preference to a normal coordinate representation since they reduce the effect of truncation [17-20]. For example, for the H_3^+ potential energy surface, Dunham [17-18] , SPF [17-19], Ogilvie [18] and various Morse expansion parameters [21] have been used for a variety of different discrete surfaces.

High-order power series often pose problems since they exhibit singularities in the domain under consideration. Hence rational functions such as Padé approximants, with an extended radii of convergence, have been used. For example, they have been used in the study of various alkali metal halides [22], Mg_2 [23], H_2 and H_2^+ [24] and LiH [25]. In general it is found that the sum of the squares of the residuals are very much smaller for Padé approximants than those obtained using power series of similar order, and so low-order expansions may be utilised, yielding fits with similar or higher precision. Furthermore, by avoiding the high-order terms, singularities may be often removed. Nevertheless, it should be noted that Murrell et al. [26] found that Padé approximants failed to provide a robust representations in a number of different applications. This still does not distract from their use as interpolating functions in the evaluation of potential energy integrals.

If the number of electronic data points are dense and extensive, then interpolation schemes using various polynomials may be more appropriate. Lagrange and Hermite interpolation schemes have been used. However, the most popular approach centres on the method of splines or fitting piecewise polynomial [27] to potential energy points. These methods have been used for interpolation of mono- and multi-dimensional ab initio potential energy surfaces [28-31].

Murrell and coworkers [3] have developed a many-body function representation for a variety of potential energy surfaces. Whitehead and Handy [32] used the Sorbie-Murrell empirical potential in a variational calculation on the vibrational band origins of H_2O in preference to a power series expansion, since the latter yielded band origins in considerable error. More recently, Vegiri and Farantos [33] have fitted an ab initio surface of HeOH based on a many-body description furnished by Murrell et al. [34].

While it is impossible to cover all the approaches to potential energy representations in this chapter, the next sub-sections deal with some established approaches namely: splines, power series (including least-squares) rational functions and many-body expansions.

§.11. Interpolating Functions.

Interpolating functions can be defined as functions which are exact at the data points or nodes. Although there is no deviation of the function at the nodes (i.e is zero) this does not mean that the quality of the fit is maintain at any intermediate point. Interpolating functions commonly used include Hermite and Lagrange functions as well as cubic splines. In order to assess the utility of these functions graphical inspections are normally undertaken and any irregularities not consistent with the physical nature of the fit is noted.

Malik et al. [28] have compared the suitability of Hermite and Lagrange functions as well as cubic splines for applications with respect to the one-dimensional Schrödinger equation.

They showed that difficulties were encountered with high-order interpolating functions when noise is introduced into the data, including that caused by truncation errors. In general, third-order expansions did not introduced irregularities, whereas fifth-order expansions introduced errors particular in the case of spline functions. The spline and Hermite interpolation schemes gave more accurate energies than Lagrangians. Spline functions have a distinct advantage over Hermite functions since knowledge of the derivatives of the potential energy at all points is not required. In particular cubic spline interpolation was found to be reliable for the representation of a "real" potential.

The use of three-dimensional cubic spline interpolation of ab initio surfaces have been further studied by Sathyamurthy and Raff [29] in the context of classical trajectory studies. A quick review of cubic splines may therefore be instructive.

Given a number of nodes (x_i, y_i) a cubic spline S of a one-dimensional potential energy surface $V(x)$ may be represented on the sub-interval $[x_i, x_{i+1}]$ by,

$$S(x) = y_i + a_i^1 (x-x_i) + a_i^2 (x-x_i)^2 + a_i^3 (x-x_i)^3 \qquad (11.1)$$

where the coefficients are given by,

$$a_i^1 = \frac{y_{i+1} - y_i}{h_i} h_i (\sigma_{i+1} + 2\sigma_i)$$

$$a_i^2 = 3\sigma_i \qquad (11.2)$$

$$a_i^3 = \frac{\sigma_{i+1} - \sigma_i}{h_i}$$

The interval size is h_i and σs are the constants which have to be determined.

Differentiating $S(x)$ three times, using the chain rule and specifying that the first and last cubics need to pass through the first four and last four data points, a system of n linear equations are obtained in n unknowns [27]. Accurate solutions of σs can be obtained using Gaussian elimination.

Cubic splines are defined in terms of fitting function values and derivatives at the nodes. Hence it does not necessarily mean that the quality of the fit is maintained at any intermediate point. The lack of precision of the interpolation often resides with the number of data points. If there are too few points, then the polynomials based on these points may introduce spurious topological structures.

Table 4.1 shows the variation of the standard deviation of spline interpolated results from the exact values for a Morse potential for H_2 as a function of the number of nodes [29]. There is a rapid improvement between the 10th and 12th node, but thereafter the improvement becomes slow varying. For a two-dimensional spline fit of the H-Cl-H surface of grid size 10x10 and 15x15 the standard deviations are 5.1×10^{-3} and 2.4×10^{-3} eV respectively [29]. For a three-dimensional spline fit of grid size 15x15x15 and 20x20x20 the standard deviations are 2.7×10^{-2} and 7.8×10^{-3} eV respectively.

Table 4.1 Standard Deviations (σ) for a Cubic Spline and a Morse Potential for H_2 as a Function of the Number of Nodes[a]

Nodes	$\sigma(V/eV)/10^{-3}$	$\Delta_1(V/eV)/10^{-3}$
10	3.1	-
12	0.8	2.3
14	0.5	0.3
16	0.2	0.3
20	0.1	0.1

a) Reproduced with permission from reference [2]. Data obtained from reference [29]. Δ_1 is just the absolute value of successive differences. The Morse function was fitted in the range $0.5 \leq R \leq 2.5$ Å.

A more testing criteria of cubic splines is their effectiveness in reproducing spectroscopic parameters. Recently, Searles and von Nagy-Felsobuki [2] fitted cubic splines to the sixteen lowest-lying states of the Li_2, using the ab initio potential energy surfaces reported by Schmidt-Mink et al. [35]. Vibrational calculations were done using 1050 finite-elements and so many more intervals than nodes were used, thereby testing the accuracy of the interpolation scheme. Table 4.2 compares the Franck-Condon factors for $1^1\Pi_u \leftarrow 1^1\Sigma_g^+$ system between theory [2, 36] and experiment [37]. The excellent agreement is evidence of both the spectroscopic quality of the discrete ab initio surface and moreover, the utility of the cubic spline as an interpolation scheme.

For one-dimensional surfaces splines are generally very accurate for a sufficient number of data points. Sathyamurthy and Raff [29] have concluded that for multi-dimensional fittings, splines are less accurate and computation times become considerable. As the number of data points required to ensure a fit of spectroscopic quality may become too large, this approach may be

Table 4.2 Franck-Condon Factors of the $1^1\Pi_u \leftarrow 1^1\Sigma_g^+$ System of Li_2[a].

v" / v'	0	1	2	3	4
0	3081 (3188)	3816 (3827)	2162 (2103)	739 (698)	170 (156)
1	3312 (3340)	111 (77)	1398 (1511)	2697 (2711)	1720 (1657)
2	2044 (2008)	839 (942)	1409 (1345)	32 (63)	1728 (1834)
3	963 (918)	1821 (1884)	10 (1)	1456 (1508)	380 (303)
4	387 (358)	1599 (1585)	610 (711)	630 (550)	557 (661)
5	141 (126)	972 (929)	1305 (1374)	6 (24)	1101 (1082)
6	48 (42)	484 (446)	1266 (1263)	531 (631)	274 (204)
7	16 (13)	214 (190)	872 (833)	1021 (1088)	31 (69)
8	5 (4)	88 (75)	496 (456)	1031 (1035)	454 (552)
9	2 (1)	35 (29)	252 (223)	772 (739)	809 (878)

v" / v'	5	6	7	8	9
0	28 (25)	3 (3)			
1	603 (560)	136 (122)	21 (18)	2 (2)	
2	2269 (2243)	1217 (1149)	375 (341)	75 (66)	10 (9)
3	521 (628)	2073 (2130)	1786 (1723)	758 (698)	196 (173)
4	1096 (1028)	3 (21)	1313 (1436)	2056 (2040)	1233 (1155)
5	12 (39)	1192 (1224)	276 (198)	471 (591)	1872 (1935)
6	814 (895)	227 (155)	663 (771)	832 (741)	21 (61)
7	744 (685)	228 (318)	713 (636)	111 (187)	1074 (1067)
8	124 (71)	780 (815)	3 (2)	835 (854)	41 (9)
9	42 (89)	505 (433)	377 (474)	277 (193)	472 (576)

a) Reproduced with permission from reference [36]. Values in brackets are experimentally determined and have been obtained from reference [37]. All entries should be multiplied by 10^{-4}.

of limited use in multi-dimensional fittings. However, coupling one-dimensional splines with other functions (such as Morse functions) have been used to circumvent these problems [38]. Moreover, in variational calculations of vibrational band origins what is required is an "overall" good fit of the potential energy surface over a defined domain. In such cases a least-squares fit to a set of points with a function(s) of desired properties is a better approach.

§.12. Power Series Expansions.

The mathematical basis for power series fits to multi-dimensional electronic surfaces is not rigorous. Hence von Nagy-Felsobuki and coworkers [2,18] have utilised the following guidelines.

(i) The power series expansion used should have a quantum mechanical basis.

(ii) Real plane convergence properties should suggest a reasonable (in a physical sense) region of acceptability.

(iii) The fit should be consistent with anticipated physical properties and so should be smooth everywhere with monotonically increasing repulsive walls.

(iv) The error of the fit should be within the estimated error associated with the ab initio calculated points.

(v) The evaluation of the expansion coefficients should be systematic and amenable to a regression analysis.

(vi) Preference should be given to an analytical representation which can accommodate several different types of experimental data.

The most common power series expansion used in multi-dimensional fits are Dunham [9-10], SPF [11] and Ogilvie [12] expansions and their exponential analogues [2, 14, 16, 35, 39-40]. It can be shown that all three satisfy criteria (i).

In general a power series expansion of the Born-Oppenheimer potential of a triatomic molecules is given by,

$$V(R_1,R_2,R_3) \approx V_0 + \sum_i^3 C_i^1 \rho_i + \sum_i^3 \sum_j^3 C_{ij}^2 \rho_i\rho_j + \sum_i^3 \sum_j^3 \sum_k^3 C_{ijk}^3 \rho_i\rho_j\rho_k$$

$$+ \sum_i^3 \sum_j^3 \sum_k^3 \sum_l^3 C_{ijkl}^4 \rho_i\rho_j\rho_k\rho_l + \cdots \qquad (12.1)$$

where ρ_i is the expansion variable and is a function of R_i and the Cs are the expansion coefficients of the fit.

The Born-Oppenheimer expansion and power series expansions, using various expansion variables noted above, are related. For example, the relationship with the SPF expansion is given by,

$$\Delta R_i = R_e \sum_{n}^{\infty} \rho_i^{(n+1)} \tag{12.2}$$

where,

$$\rho_i = \Delta R_i / R_i = (R_i - R_e)/R_i \tag{12.3}$$

The Dunham, SPF and Ogilvie expansions are related via,

$$\rho_i^{OGILVIE} = \frac{2 \rho_i^{SPF}}{2 - \rho_i^{SPF}} = \frac{2 \rho_i^{DUNHAM}}{2 + \rho_i^{DUNHAM}} = \frac{2(R_i - R_e)}{R_i + R_e} \tag{12.4}$$

Furthermore, Thakkar [13] has shown that the Morse-type expansion variable (i.e. $\{1-\exp(-a(R-R_e)\})$ also has a quantum mechanical basis. Hence all power series expansion discussed below satisfy criterion (i). Further, both the Dunham, SPF and Ogilvie force fields have been used to take into account several types of experimental data, thereby satisfying criteria (vi).

Criterion (ii) is more problematic since real plane convergence properties are not only dependent on the expansion parameter used in the power series fit, but also on the electronic potential energy surface. However, such studies are tractable for one-dimensional problems. In the case of the Dunham expansion, Beckel and Engelke [41] have shown that because of the united atom pole its domain of convergence is limited to $0 < R_i < 2R_e$. Consequently, Dunham potentials are generally unsuitable for computing properties with large R (for example, eigenvalues of the high vibrational levels near the dissociation limit). On the other hand, the SPF expansion radius of convergence is $R_i > 0.5R_e$ [42]. Although this expansion extends the region of applicability, there has been observed oscillatory behaviour in diatomic potentials [11] that originate from sub-domain outside the real plane radius of convergence. Of the three, only the Ogilvie power series expansion may be valid over the entire range [12]. However, Beckel [42] has shown, using a stylised potential, that the Ogilvie expansion may be more restricted by singularities at $R = 0$ and as $R \rightarrow \infty$ than either the Dunham or SPF expansion.

The convergence properties of the Morse-type expansion have long been recognised [13, 16], although infrequently used. As Huffaker [14] has pointed out, the Morse expansion parameter itself is a good approximation to the potential of a diatomic molecule and so the convergence properties should be good for perturbed Morse potentials. It is therefore not surprising that the more recent power series representations of triatomic molecular potentials have centred on the exponential variants [2, 21, 39-40].

Criterion (v) requires the assumption that the molecular functional form V is well behaved and amenable to a χ^2 analysis. Hence, to determine a power series fit to the discrete surface the sum of the squares of the residuals needs to be minimised. It is given by,

$$L = \sum_{m=1}^{N} \{ V^{ps}(\rho_{1m}, \rho_{2m}, \rho_{3m}) - V^{exact}(\rho_{1m}, \rho_{2m}, \rho_{3m}) \}^2 \qquad (12.5)$$

where N is the number of potential data points, V^{ps} is the potential determined using the power series expansion and V^{exact} is the exact potential at that point. An estimate of the expansion coefficients as well as an estimate of the fitted error, χ^2 is therefore obtained.

The power series with the smallest χ^2 are not necessarily the best fit to a surface. Criterion (iii) specifies that the function should be consistent with anticipated physical properties. High-order power series often pose problems with singularities in the domain under consideration. The singular value decomposition technique is an effective method which enables the magnitude of high-order coefficients to be appropriately dampened.

The SVD analysis as described by Wilkinson [43] and Forsythe et al. [27]. It is used by von Nagy-Felsobuki and coworkers [2, 39-40, 44-45] to identify near rank deficiencies in linear least squares fits of multi-dimensional surfaces.

Writing the residual as,

$$R(\alpha) = (\mathbf{b} - A\,\alpha)^T(\mathbf{b} - A\,\alpha) \qquad (12.6)$$

where A is a matrix of rank m x n (m >n) and the superscript T denotes the transpose. The A matrix can be factorised into,

$$A = U\Sigma V^T \qquad (12.7)$$

where U is an orthogonal matrix of rank m x m, V is an n x n orthogonal matrix and Σ is a m x n matrix with entries σ_i ($1 \leq i \leq n$) on the main diagonal and all other entries zero.

From the orthogonal properties of U and V, $R(\alpha)$ can be rewritten as,

$$R(\alpha) = (\mathbf{c} - \Sigma\mathbf{d})^T(\mathbf{c} - \Sigma\mathbf{d}) \tag{12.8}$$

where $\mathbf{d} = V^T\alpha$ and $\mathbf{c} = U^T\mathbf{b}$.

The minimum residual is given by,

$$R^* = \sum_j c_j^2 \tag{12.9}$$

obtained using the solution vector,

$$\alpha = V\mathbf{d} \tag{12.10}$$

with $d_i = c_i/s_i$, $1 \le i \le n$ and $\|\alpha\|^2 = \alpha^T\alpha = \sum_i (c_i/s_i)^2$. $\tag{12.11}$

Treating any σ_i as zero increases the residual,

$$R \leftarrow R^* + c_i^2 \tag{12.12}$$

and reduces the norm of the solution vector,

$$\|\alpha\|^2 \leftarrow \|\alpha\|^2 - (c_i/\sigma_i)^2 \tag{12.13}$$

Entries i for which c_i is relatively small but for which (c_i/σ_i) is relatively large require special attention. The corresponding column of V^T is a linear combination of parameters which is poorly determined by the data. Setting $\sigma_i = 0$ will not appreciably degrade the fit as measured by function $R(\alpha)$, but will substantially reduce $\|\alpha\|$ which is a desirable outcome for the high-order terms of a Taylor series expansion.

At this point it would be instructive to illustrate the above concepts with a multi-dimensional surface. The discrete ab initio surfaces of LiH_2^+ developed by Searles and von Nagy-Felsobuki [40] is a good example of the difficulties encountered in power series fit of a four dimensional problem. As the molecule possess C_{2v} symmetry the energy hypersurface can be described in the t coordinate system [39]. The power series expansions of Dunham, SPF and Ogilvie and the Morse-type variants were examined in detail [40].

Of all the possible power series variants investigated, the Morse-Dunham and Ogilvie expansions most accurately reproduce the LiH_2^+ discrete surface. The exponential Dunham expansion variable is given by,

$$\rho_i = 1-\exp\{-a(R_i-R_e)/R_e\} \qquad\qquad (12.14)$$

where $a = 1$ and the Ogilvie expansion variable is given by equation (12.4).

The coefficients calculated for the optimum fits are given in Table 4.3 with the 6th order exponential Dunham and the Ogilvie expansion variables giving the smallest sum of squares of residuals. In particular, the χ^2 obtained using the 6th order fits are 1.4×10^{-4} au^2 and 4.5×10^{-5} au^2 respectively. It should be noted that the high-order coefficients are large in magnitude, and this is particularly the case in the exponential Dunham fit.

Figures 4.1(a)-(c) show energy contour plots using the 6th order Ogilvie fit and Figures 4.2 (a)-(c) show energy contour plots using the 6th order exponential Dunham fit. In all figures the energy contours represents an increment of 50 kJ mol^{-1}. Unlike the 6th order Ogilvie fit, graphical examination of the physical nature of the 6th order exponential Dunham fit indicates that in the region defined by the data the function does not satisfy the criteria of being smooth everywhere with monotonically increasing repulsive walls (see criterion (iii)). Some of the high-order coefficients for both fits are unreasonably large, especially since the expectation is that in a convergent Taylor series expansion the coefficients should dampen. An SVD analysis of the Ogilvie fit suggests that σ_{47} should be set to zero. That is, the SVD analysis identifies that this expansion variable is having a large influence of the size of the coefficients, but a negligible effect on the χ^2. As is highlighted in Table 4.3 removal of this expansion variable from the fitting procedure reduces the magnitude of the larger high-order coefficients and yields a χ^2 of 4.8×10^{-5} au^2, which has only marginally degraded the fit.

A similar SVD analysis of the 6th order exponential Dunham fit did not eliminate singularities unless eleven (σ_{25}, σ_{31}, σ_{37} and σ_{41}-σ_{48}) of the expansion variables were removed. The resulting fit gave a χ^2 value of 7.8×10^{-4} au^2, which is only slightly better than the 5th order exponential Dunham fit, which was smooth without SVD analysis. The χ^2 of the fits in Table 4.3 are too poor to be appropriate for rovibrational or scattering calculations. Consequently, investigations were conducted into use of Padé approximants [40].

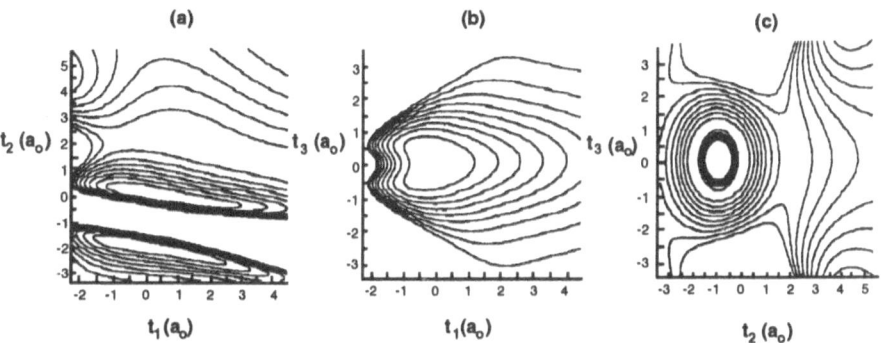

Figure 4.1 Contour plots for the LiH_2^+ potential-energy surface using the 6th order Ogilvie fit. Contours are in increments of 50 kJmol^{-1}. (a) t_1 versus t_2 (b) t_1 versus t_3 (c) t_2 versus t_3. Reproduced with permission from reference [40].

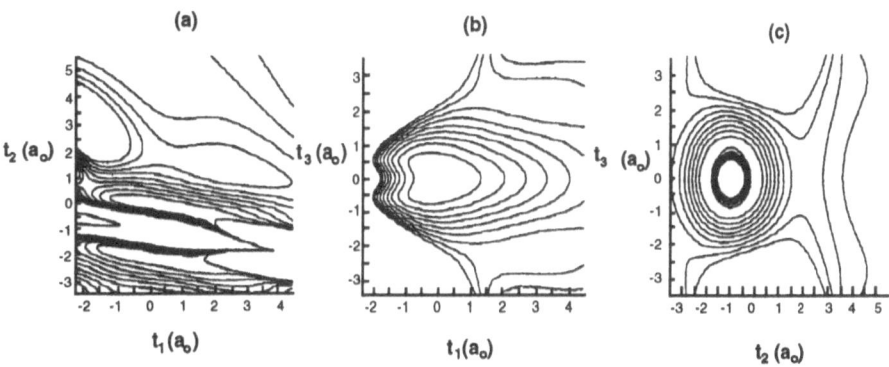

Figure 4.2 Contour plots for the LiH_2^+ potential-energy surface using the 6th order exponential Dunham fit. Contours are in increments of 50 kJmol^{-1}. (a) t_1 versus t_2 (b) t_1 versus t_3 (c) t_2 versus t_3. Reproduced with permission from reference [40].

Table 4.3 Expansion Coefficients for Power Series Expansion of the LiH_2^+ Potential Energy Surface[a].

	Expansion Variable	Expansion Coefficient		
		Morse-Dunham 6th [b]	Ogilvie 6th [c]	Ogilvie 6th [d]
	1	-8.434844	-8.435093	-8.435128
1	$\rho_1 + \rho_2$	0.000000	0.000000	0.000000
2	ρ_3	0.000000	0.000000	0.000000
3	$\rho_1^2 + \rho_2^2$	0.094844	0.087905	0.098883
4	ρ_3^2	0.336435	0.349776	0.348998
5	$\rho_1\rho_2$	-0.081554	-0.078844	-0.105227
6	$\rho_2\rho_3 + \rho_1\rho_3$	-0.010715	0.001400	0.004434
7	$\rho_1^3 + \rho_2^3$	-0.086444	-0.101069	-0.100206
8	ρ_3^3	-0.226558	-0.204671	-0.204772
9	$\rho_1^2\rho_2 + \rho_2^2\rho_1$	-0.065493	-0.074700	-0.068586
10	$\rho_1^2\rho_3 + \rho_2^2\rho_3$	-0.005163	0.000355	-0.004215
11	$\rho_1\rho_3^2 + \rho_2\rho_3^2$	-0.018884	0.017275	0.016599
12	$\rho_1\rho_2\rho_3$	0.134048	0.153850	0.150419
13	$\rho_1^4 + \rho_2^4$	0.171853	0.064532	-0.027821
14	ρ_3^4	0.241578	0.089549	0.092963
15	$\rho_1^3\rho_2 + \rho_2^3\rho_1$	0.072841	-0.233736	0.191805
16	$\rho_1^3\rho_3 + \rho_2^3\rho_3$	-0.141718	-0.123296	-0.142449
17	$\rho_1\rho_3^3 + \rho_2\rho_3^3$	0.133445	-0.062103	-0.082659
18	$\rho_1^2\rho_2^2$	-0.158131	0.654179	0.035417

Table 4.3 (Cont.)

19	$\rho_1^2\rho_3^2 + \rho_2^2\rho_3^2$	-0.045421	0.114928	0.116776
20	$\rho_1^2\rho_2\rho_3 + \rho_1\rho_2^2\rho_3$	0.137495	-0.217777	-0.239985
21	$\rho_1\rho_2\rho_3^2$	-0.240227	0.326486	0.390608
22	$\rho_1^5 + \rho_2^5$	0.017230	-0.080993	-0.170827
23	ρ_3^5	-0.079732	-0.098022	-0.098100
24	$\rho_1^4\rho_2 + \rho_2^4\rho_1$	0.524208	-0.062582	0.371860
25	$\rho_1^4\rho_3 + \rho_2^4\rho_3$	-0.297709	-0.190190	-0.063139
26	$\rho_1\rho_3^4 + \rho_2\rho_3^4$	0.076903	-0.117500	-0.116784
27	$\rho_1^3\rho_2^2 + \rho_1^2\rho_2^3$	-0.389916	0.539347	0.129610
28	$\rho_1^3\rho_3^2 + \rho_2^3\rho_3^2$	-0.056254	0.106929	0.221474
29	$\rho_1^2\rho_3^3 + \rho_2^2\rho_3^3$	0.248042	0.010127	0.005617
30	$\rho_1^3\rho_2\rho_3 + \rho_1\rho_2^3\rho_3$	-0.099259	0.517516	-0.400955
31	$\rho_1\rho_2\rho_3^3$	0.848577	-0.650824	-0.408477
32	$\rho_1^2\rho_2^2\rho_3$	1.442121	-2.571352	-0.551895
33	$\rho_1^2\rho_2\rho_3^2 + \rho_1 r_2^2\rho_3^2$	-1.109406	0.792642	0.408204
34	$\rho_1^6 + \rho_2^6$	-0.088730	0.125692	0.142431
35	ρ_3^6	-0.114397	0.006176	0.003416
36	$\rho_1^5\rho_2 + \rho_2^5\rho_1$	1.089830	-0.347578	-0.429630
37	$\rho_1^5\rho_3 + \rho_2^5\rho_3$	-0.628374	0.261829	0.387655

Table 4.3 (Cont.)

38	$\rho_1\rho_3^5 + \rho_2\rho_3^5$	-0.455841	0.174704	0.195924
39	$\rho_1^4\rho_2^2 + \rho_2^4\rho_1^2$	-0.167977	0.301366	0.631478
40	$\rho_1^4\rho_3^2 + \rho_2^4\rho_3^2$	0.008483	0.250191	0.204528
41	$\rho_1^2\rho_3^4 + \rho_2^2\rho_3^4$	-0.002332	-0.176176	-0.185188
42	$\rho_1^4\rho_2\rho_3 + \rho_2^4\rho_1\rho_3$	0.326934	-0.253277	0.841289
43	$\rho_1\rho_2\rho_3^4$	-0.035357	0.155178	-0.177458
44	$\rho_1^3\rho_2^3$	-2.000826	0.620968	-0.062151
45	$\rho_1^3\rho_3^3 + \rho_2^3\rho_3^3$	0.651458	-0.103507	-0.209476
46	$\rho_1^3\rho_2^2\rho_3 + \rho_1^2\rho_2^3\rho_3$	2.096800	-0.950732	-0.170213
47	$\rho_1^3\rho_2\rho_3^2 + \rho_1\rho_2^3\rho_3^2$	-1.433352	-0.126393	0.348465
48	$\rho_1^2\rho_2\rho_3^3 + \rho_1\rho_2^2\rho_3^3$	1.144989	-0.397705	0.157739
49	$\rho_1^2\rho_2^2\rho_3^2$	-3.318453	2.063729	0.140001
χ^2		1.4×10^{-4}	4.5×10^{-5}	4.8×10^{-5}

a) All entries in atomic units.

b) Sixth-order Morse-Dunham expansion.

c) Sixth-order Ogilvie expansion.

d) Sixth-order Ogilvie expansion with SVD analysis (σ_{47} set to zero).

§.13. Rational Functions.

Rational functions (or Padé approximants) are functions in which the numerator and denominator are power series expansions (of order m and n respectively) of variables ρ_i (which are functions of R_i). That is, they can be written as,

$$P(m,n) = \frac{\sum_{i=0}^{m} \sum_{j=0}^{m} \sum_{k=0}^{m} a_{ijk} \rho_1^i \rho_2^j \rho_3^k}{\sum_{i'=0}^{n} \sum_{j'=0}^{n} \sum_{k'=0}^{n} b_{i'j'k'} \rho_1^{i'} \rho_2^{j'} \rho_3^{k'}} \qquad (13.1)$$

such that $(i + j + k) \le m$ and $(i' + j' + k') \le n$.

Murrell et al. [26] has found that the rational function expansion failed in yielding a robust representation of diatomic potentials for a wide range of applications. Singularities were found in the attractive and the repulsive region of the curve (the position of these singularities being unpredictable and varying with the order of the power series in the numerator and denominator). Murrell et al. [26] also investigated the introduction of a $(R_i)^{-1}$ factor into the function and discovered that the resulting function gave a more satisfactory representation of the systems studied, even though singularities were still apparent.

The application of Padé approximants to multi-dimensional surfaces remains largely unexplored, due to possible inherent problems with singularities (since the power series in the denominator may equal zero at geometries where this is physically unacceptable). However, Padé approximants are much more flexible than power series expansions and therefore may be necessary where power series expansions fail to produce an adequate and precise representation of the potential energy surface. This was found to be the case for the discrete surface of LiH_2^+ [40].

In the case of LiH_2^+ [40] the six variables previously used for the power series expansions were also used as expansion variables in the Padé approximant representations. In each case, the χ^2 of the ensuing fits are lower than those for the respective power series representations. However, on graphical examinations it was revealed that the Padé surfaces suffered from a substantial increase in the number of singularities. In addition, the singularities are not removed by use of the SVD analysis. For example, Figures 4.3 (a)-(c) give the energy contour plots with respect to t coordinates for a 6th order Padé approximate (denoted P(6,6)) using Dunham expansion variables. Of all the variants, this fit gives the lowest χ^2, 1.1×10^{-8} au^2.

An alternative form of the Padé approximant was tested to represent the LiH_2^+ surface. It has the form [40],

$$P'(m,n) = \frac{\sum\limits_{i=0}^{m} \sum\limits_{j=0}^{m} \sum\limits_{k=0}^{m} a_{ijk}\rho_1^i\rho_2^j\rho_3^k}{R_1 R_2 R_3 \sum\limits_{i'=0}^{n} \sum\limits_{j'=0}^{n} \sum\limits_{k'=0}^{n} b_{i'j'k'}\rho_1^{i'}\rho_2^{j'}\rho_3^{k'}} \qquad (13.2)$$

such that $(i + j + k) \leq m$ and $(i' + j' + k') \leq n$.

The two R_{Li-H} separations are represented by R_1 and R_2 and the R_{H-H} separation by R_3. This function gives a small χ^2 for all the variants, but in general the high-order expansion coefficients are large and graphical inspections of the contour plots reveal many singularities. However, the Dunham expansion which is 6th order in the numerator and 4th order in the denominator (that is, P'(6,4)), gives a function which is smooth and has monotonically increasing repulsive walls over the region defined by the data points. The fit gives a χ^2 of 1.3×10^{-7} au^2 which indicates that this fit is substantially more accurate than the best power series expansion. Contour plots for this surface are given in Figures 4.4 (a)-(c) and the expansion coefficients of the P'(6,4) Padé approximate are given in Table 4.4.

The calculation and analysis of the LiH_2^+ [40] potential energy surface indicates that although the power series expansions with SVD analysis could be used to give a smooth analytical representation of the potential energy surface, they yield a large χ^2. The LiH_2^+ surface contains many features for which the power series expansions are too rigid to accommodate. Padé approximants generally gave better χ^2 values. Except for the P'(6,4) expansion, many singularities occurred in the domain of interest. The smoothness of the P'(6,4) expansion was unpredictable and rational functions of similar order are not smooth. The study indicated that the power series expansion would probably be more appropriate for molecules whose potential is quite regular. However, for irregular surfaces it is necessary to use a Padé.

§.14. Many-Body Expansions.

Murrell and coworkers [3, 34, 46-48] have reactivated many-body expansions of potential energy surfaces. The London, Eyring, Polanyi and Sato function [49] and the diatomics-in-molecules approach [50] can be thought of being in this class. In the case of triatomic molecules, Murrell and coworkers [3] have successfully applied their many-body function approach to a variety of systems. It is therefore not surprising that their functional representation is used by a number of other workers and that variations of their approach are appearing in the literature [33].

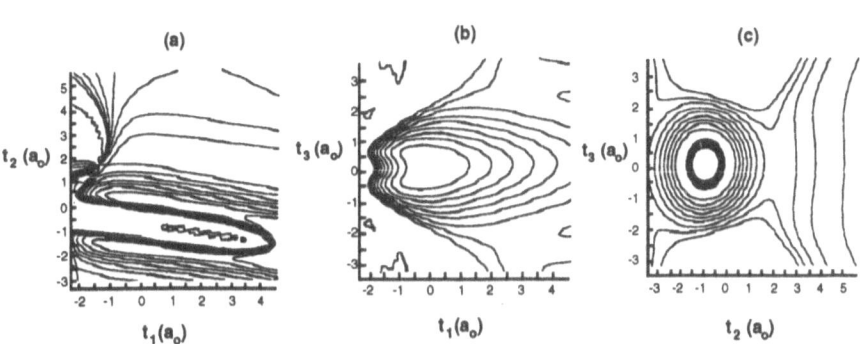

Figure 4.3 Contour plots for the LiH_2^+ potential-energy surface using the P(6,6) Padé-approximant fit, with Dunham expansion variables. Contours are in increments of 50 kJmol⁻¹. (a) t_1 versus t_2 (b) t_1 versus t_3 (c) t_2 versus t_3. Reproduced with permission from reference [40].

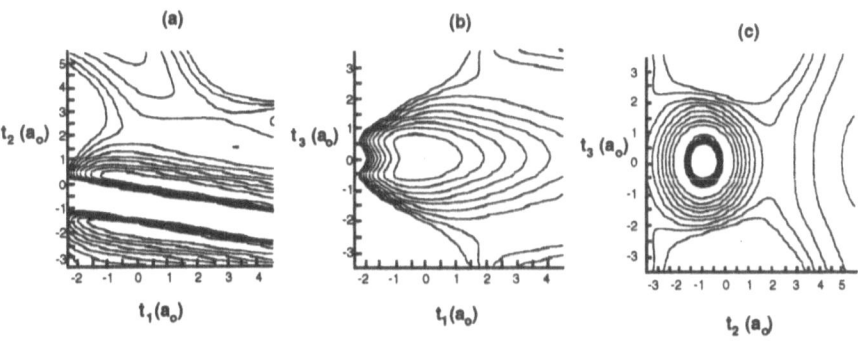

Figure 4.4 Contour plots for the LiH_2^+ potential-energy surface using the P'(6,4) Padé-approximant fit with Dunham expansion variables. Contours are in increments of 50 kJmol⁻¹. (a) t_1 versus t_2 (b) t_1 versus t_3 (c) t_2 versus t_3. Reproduced with permission from reference [40].

Table 4.4 Expansion Coefficients for Dunham Padé Approximate Expansion of the LiH_2^+ Potential Energy Surface[a].

	Expansion Variable	Expansion Coefficient Numerator	Denominator
	1	-178.2047	1.0000
1	$\rho_1 + \rho_2$	-336.4846	0.88819
2	ρ_3	-161.7542	-0.09231
3	$\rho_1^2 + \rho_2^2$	-208.6695	0.29381
4	ρ_3^2	41.8770	-0.10078
5	$\rho_1\rho_2$	-695.5297	1.11638
6	$\rho_2\rho_3 + \rho_1\rho_3$	-292.5161	-0.15511
7	$\rho_1^3 + \rho_2^3$	-94.2672	0.23376
8	ρ_3^3	2.0409	0.05972
9	$\rho_1^2\rho_2 + \rho_2^2\rho_1$	-354.4265	-0.30668
10	$\rho_1^2\rho_3 + \rho_2^2\rho_3$	-157.0927	-0.13708
11	$\rho_1\rho_3^2 + \rho_2\rho_3^2$	98.3250	-0.12534
12	$\rho_1\rho_2\rho_3$	-681.6319	0.33684
13	$\rho_1^4 + \rho_2^4$	-43.3469	0.01744
14	ρ_3^4	-8.8394	-0.00066
15	$\rho_1^3\rho_2 + \rho_2^3\rho_1$	-39.7770	-0.00052
16	$\rho_1^3\rho_3 + \rho_2^3\rho_3$	-66.8614	-0.02461
17	$\rho_1\rho_3^3 + \rho_2\rho_3^3$	18.4490	0.00295
18	$\rho_1^2\rho_2^2$	-89.4748	-0.00443

Table 4.4 (Cont.)

19	$\rho_1^2\rho_3^2 + \rho_2^2\rho_3^2$	81.1499	0.00874
20	$\rho_1^2\rho_2\rho_3 + \rho_1\rho_2^2\rho_3$	-360.3963	-0.00258
21	$\rho_1\rho_2\rho_3^2$	106.3896	-0.00259
22	$\rho_1^5 + \rho_2^5$	-2.9803	-
23	ρ_3^5	-0.1271	-
24	$\rho_1^4\rho_2 + \rho_2^4\rho_1$	-46.0157	-
25	$\rho_1^4\rho_3 + \rho_2^4\rho_3$	-38.8884	-
26	$\rho_1\rho_3^4 + \rho_2\rho_3^4$	-7.8769	-
27	$\rho_1^3\rho_2^2 + \rho_1^2\rho_2^3$	56.0775	-
28	$\rho_1^3\rho_3^2 + \rho_2^3\rho_3^2$	30.0708	-
29	$\rho_1^2\rho_3^3 + \rho_2^2\rho_3^3$	13.0266	-
30	$\rho_1^3\rho_2\rho_3 + \rho_1\rho_2^3\rho_3$	-12.3412	-
31	$\rho_1\rho_2\rho_3^3$	30.9005	-
32	$\rho_1^2\rho_2^2\rho_3$	-145.8230	-
33	$\rho_1^2\rho_2\rho_3^2 + \rho_1 r_2^2\rho_3^2$	30.9203	-
34	$\rho_1^6 + \rho_2^6$	-1.5057	-
35	ρ_3^6	0.0213	-
36	$\rho_1^5\rho_2 + \rho_2^5\rho_1$	-0.8166	-
37	$\rho_1^5\rho_3 + \rho_2^5\rho_3$	-2.3367	-

Table 4.4 (Cont.)

38	$\rho_1\rho_3^5 + \rho_2\rho_3^5$	-0.3036	-
39	$\rho_1^4\rho_2^2 + \rho_2^4\rho_1^2$	-2.4217	-
40	$\rho_1^4\rho_3^2 + \rho_2^4\rho_3^2$	4.4660	-
41	$\rho_1^2\rho_3^4 + \rho_2^2\rho_3^4$	1.2108	-
42	$\rho_1^4\rho_2\rho_3 + \rho_2^4\rho_1\rho_3$	-37.2704	-
43	$\rho_1\rho_2\rho_3^4$	-7.0436	-
44	$\rho_1^3\rho_2^3$	5.8807	-
45	$\rho_1^3\rho_3^3 + \rho_2^3\rho_3^3$	-3.2093	-
46	$\rho_1^3\rho_2^2\rho_3 + \rho_1^2\rho_2^3\rho_3$	51.3713	-
47	$\rho_1^3\rho_2\rho_3^2 + \rho_1\rho_2^3\rho_3^2$	28.0094	-
48	$\rho_1^2\rho_2\rho_3^3 + \rho_1\rho_2^2\rho_3^3$	11.7577	-
49	$\rho_1^2\rho_2^2\rho_3^2$	-51.1058	-

χ^2 $\qquad\qquad\qquad\qquad\qquad\qquad\qquad\qquad$ 1.3×10^{-7}

---

a) Reproduced with permission from reference [40]. All entries in atomic units.

An important aspect of the many-body expansion is that it is a flexible global representation of a surface. The formulations of Murrell and coworkers [3] contain a sufficient number of parameters in order that the final form of the functions can accurately reproduce the energy, geometry, and forces constants at the equilibrium geometry. In the case of a triatomic, the function takes on the form of an atom-diatom representation and so must be engineered to reproduce the dissociation energy and diatomic potential energy curves of all possible atom-diatom products. Of course, if fitting to spectroscopic data all these quantities must be predetermined. In the case of discrete ab initio potential energy surfaces, problems are encountered due to the lack of accuracy near the two body asymptotic limits. Murrell and coworkers [3] have replaced the ab initio one-body and two-body terms in the potential by empirical terms.

Murrell and coworkers [3] have given a thorough description of their many-body formulations. Nevertheless, it might be informative to highlight their approach by outlining the Sorbie-Murrell [48] empirical potential. For a triatomic molecule it has form,

$$V_{XYZ}(R_1,R_2,R_3) = V_{XY}(R_1) + V_{YZ}(R_2) + V_{ZX}(R_3) + V_{XYZ}(R_1,R_2,R_3) \qquad (14.1)$$

where R_i $(i = 1,2,3)$ is the internuclear distance between each pair of labelled atoms and V_{ij} $(i,j = X,Y, Z)$ represents a diatomic potential formed by adiabatic dissociation of the molecule XYZ.

The three-body terms in the expansion has the form,

$$V_{XYZ}(R_1,R_2,R_3) = \sum_{ijk} C_{ijk}\, s_1^i s_2^j s_3^k \prod_m (1\text{-tanh}\,(\gamma_m s_m/2)) \qquad (14.2)$$

where s_m are the displacement co-ordinates and the values for the coefficients C_{ijk} and γ_m are obtained from experiment and from other data for molecules studied.

With respect to SO_2 Murrell and coworkers [34, 46-47] have published three potential functions. The first labelled Pot 1 [46] was based on experimental properties of the equilibrium configuration possessing C_{2v} symmetry and on ab initio SCF calculations of the energy and geometry of a metastable minimum of C_s. The second potential labelled Pot 2 [34] used the spectroscopic, thermochemical and ab initio data of Pot 1, but with the potential parameters varied to minimise the difference between the experimental vibrational frequencies and those calculated by the potential using a variation method. Unfortunately Pot 1 & 2 give a barrier which increases from the collinear to C_{2v} approaches (which is contrary to experiment). The third potential labelled Pot 3 [47] was obtained by treating the coefficients as linear parameters for selected γ_i values where the optimum γ_i values were chosen by imposing the criteria that there should be no barrier for approach angle SOO > 140° and no barrier for the removal of O from SO_2 along paths close to the equilibrium angle. However, there would be a barrier for 90° < SOO < 140°.

Table 4.5 compares the experimentally [51] and the theoretically derived force constants using Pot 1 & Pot 2. The force constants are calculated from the respective potentials by differentiation at the equilibrium geometry with initial input in the optimisation procedure for both potentials requiring the experimentally derived harmonic forces constants, which are then further refined. It is clear that Pot 1 deviates significantly from the Strey-Mills force field [51], whereas Pot 2 essentially reproduces the experimentally derived result.

Table 4.6 compares the experimental [52] and theoretical vibrational frequencies [34, 47], the latter being calculated using Pot 2 and Pot 3 in the Watson normal coordinate Hamiltonian [53] and adopting the variational procedure of Whitehead-Handy [54] with 56 basis functions being employed. Obviously near equilibrium both potentials are of similar quality, although Pot 3 has a better representation of barriers in the entrance channel of the $S(^3P)$ and $O_2(^3\Sigma_g^-)$ reaction.

Table 4.5 Experimental and Theoretical Force Constants of $SO_2{}^a$.

	Exp[b]	Pot 1[c]	Pot 2[d]
f_{11}	10.420	6.6	10.410
f_{22}	10.420	12.4	10.410
f_{12}	0.124	5.0	0.120
$f_{\alpha\alpha}$	1.680	1.04	1.640
$f_{1\alpha}$	0.525	1.41	0.563
$f_{2\alpha}$	0.525	0.57	0.563

a) Reproduced with permission from reference [2]. Units are as follows: $f_{11} = f_{22} = f_{12} = aJ\text{Å}^{-2}$; $f_{1\alpha} = f_{2\alpha} = aJ\text{Å}^{-2}$; $f_{\alpha\alpha} = 1.04$ aJ.
b) Data obtained from reference [51].
c) Data obtained from reference [46].
d) Data obtained from reference [34].

Table 4.6 Experimental and Theoretical Vibration Band Origins of SO_2[a] (in cm^{-1}).

Level	Exp[b]	Pot 1[c]	Pot 2[d]
100	1151	1151	1154
010	518	515	515
001	1362	1362	1367

a) Reproduced with permission from reference [2].

b) Data obtained from reference [52].

c) Data obtained from reference [34].

d) Data obtained from reference [47].

REFERENCES TO CHAPTER IV

1 Brown WB, Steiner E (1966) J Chem Phys 44:3934

2 Searles DJ, von Nagy-Felsobuki EI (1991) In: Vibrational spectra and structure, Durig JR (Ed), Vol 19, Elsevier, Amsterdam

3 Murrell JN, Carter S, Farantos SC, Huxley P, Varandas AJC (1984) Molecular potential energy functions, John Wiley & Sons, Brisbane

4 Mezey PG (1987) Potential energy hypersurfaces, Elsevier, Amsterdam

5 Varandas AJC (1990) In: Trends in atomic and molecular physics, Univeridad, Yanez M (Ed), Autonomade Madrid, Madrid

6 Le Roy RJ (1980) In: Semiclassical methods in molecular scattering and spectroscopy, Child MS (Ed), Reidel, London

7 Vidal CR, Scheingraber H (1977) J Mol Spectrosc 65:46

8 Pardo A, Poyato JML, Camacho JJ, Martin E (1988) Spectrochim Acta Part A44:335

9 Dunham JL (1932) Phys Rev 41:713

10 Dunham JL (1932) Phys Rev 41:721

11 Simons G, Parr RG, Finlan JM (1973) J Chem Phys 59:3229

12 Ogilvie JF (1981) Proc R Soc London Ser A378:287

13 Thakkar AJ (1975) J Chem Phys 62:1693

14 Huffaker JN (1976) J Chem Phys 64:3175

15 Morse PM (1929) Phys Rev 34:57

16 Coolidge AS, James HM, Vernon EL (1938) Phys Rev 54:726

17 Carney GD, Porter RN (1976) J Chem Phys 65:3547

18 Burton PG, von Nagy-Felsobuki EI, Doherty G, Hamilton M (1985) Mol Phys 55:527

19 Burton PG, von Nagy-Felsobuki EI, Doherty G, Hamilton M (1984) Chem Phys 83:83

20 Carney GD, Curtiss LA, Langhoff SR (1976) J Mol Spectrosc 61:371

21 Meyer W, Botschwina P, Burton PG (1986) J Chem Phys 84:891

22 Jordan KD, Kinsey JL, Silbey R (1974) J Chem Phys 61:911

23 Jorish VS, Scherbak NB (1979) Chem Phys Lett 67:160

24 Sonnleitner SA, Beckel CL, Colucci AJ, Scaggs ER (1981) J Chem Phys 75:2018

25 Pardo A, Camacho JJ, Poyato JML (1986) Chem Phys Lett 131:490

26 Murrell JN, Varandas AJC, Brandao J (1987) Theor Chim Acta 71:459

27 Forsythe GE, Malcolm MA, Moler CB (1977) Computer methods for mathematical computations, Prentice-Hall, New York

28 Malik DJ, Eccles J, Secrest D (1980) J Comput Phys 38:157

29 Sathyamurthy N, Raff LM (1975) J Chem Phys 63:464

30 Dunne SJ, Searles DJ, von Nagy-Felsobuki EI (1987) Spectrochim Acta A43: 699

31 Bruehl M, Schatz GC (1988) J Phys Chem 92:7223

32 Whitehead RJ, Handy NC (1976) J Mol Spectrosc 59:459

33 Vegiri A, Farantos SC (1988) J Phys Chem 92:2723

34 Carter S, Mills IM, Murrell JN, Varandas AJC (1982) Mol Phys 45:1053

35 Schmidt-Mink I, Müller W, Meyer W (1985) Chem Phys 92:263

36 Searles DJ, Burke T, von Nagy-Felsobuki EI (1988) AINSE Technical Report No. 88/21.

37 Hessel MM, Vidal CR (1979) J Chem Phys 70:4439

38 Bowman JM, Kuppermann A (1975) Chem Phys Lett 34:523

39 Searles DJ, Dunne SJ, von Nagy-Felsobuki EI (1988) Spectrochim Acta A44:505

40 Searles DJ, von Nagy-Felsobuki EI (1991) Phys Rev A43:3365

41 Beckel CL, Engelke R (1968) J Chem Phys 49:5199

42 Beckel CL (1976) J Chem Phys 65:4319

43 Wilkinson JH (1978) In: Numerical software: needs and availability, Jacobs D (Ed), Academic Press, New York

44 Searles DJ, von Nagy-Felsobuki EI (1992) Comp Phys Comm 67:527

45 Wang F, Searles DJ, von Nagy-Felsobuki EI (1992) J Chin Chem Soc 39:339

46 Farantos S, Leisegang EC, Murrell JN, Sorbie K, Texeira-Dias JJC, Varandas AJC (1977) Mol Phys 34:947.

47 Murrell JN, Craven W, Vincent M, Zhu ZH (1985) Mol Phys 56:839

48 Sorbie KS, Murrell JN (1975) Mol Phys 29:1387

49 Sato S (1955) J Chem Phys 23:2465

50 Ellison FO (1963) J Am Chem Soc 85:3540.

51 Mills IM (1974) Theoretical Chemistry, Chemical Society Specialist Periodical Report,Vol. 10, London

52 Shelton RD, Nielson HH, Fletcher WH (1954) J Chem Phys 22:1791

53 Watson JKG (1968) Mol Phys 15:479

54 Whitehead RJ, Handy NC (1975) J Mol Spectrosc 55:356

3. Watson JKG [J. Mol. Spectrosc. 1984, 103, ...]

4. Strickland RJ, Hervey J. Chem. Phys. 21 J. Mol. Spectrosc. 65, 3, 70

CHAPTER V

FINITE-ELEMENT SOLUTION OF ONE-DIMENSIONAL SCHRÖDINGER EQUATIONS

§.15. Numerical Solutions of One-Dimensional Schrödinger Problems.

Analytical solutions for the one-dimensional Schrödinger problems can only be obtained for contrived potential energy functions such as the finite-square well, the simple harmonic oscillator and Morse potential problems. As the eigenenergies and eigenfunctions of these systems are known exactly, they serve as useful systems for the assessment of solution algorithms to be applied to more general problems.

In general the Schrödinger equation cannot be solved analytically. Traditionally, chemists and physicists have solved the Schrödinger equation using spectral methods, where a finite global basis set is used to give a global solution to the problem. The basis set often comprises functions which are solutions to analytically solvable problems with Hamiltonians closely related to the Hamiltonian under investigation. This approach is insufficiently flexible for potentials which have localised features (unless a large number and variety of basis functions are used).

There are numerous algorithms (and variants) for solving the one-dimensional nuclear Schrödinger equation, some of which have been compared by Shore [1], Malik et al. [2] and Duff et al. [3]. The Numerov-Cooley method [4-5] and methods obtained from its modification [6-12] are "shooting" methods. That is, an initial approximate is specified for the eigenenergy which is then used in the iterative computation of the eigenenergy to the desired level of accuracy. Eigenfunctions are determined singly using trial eigenenergies. The vibrational quantum number is verified by counting the number of nodes in the eigenfunction. The method is easily programmed and does not require large amounts of memory. Therefore, it has been widely used for the solution of the one-dimensional Schrödinger equation. The accuracy of the eigenfunctions is of order $O(h^4 n^6)$ [13-14], where h is the spacing of grid points used in the calculation and n specifies the energy level. Hence, the precision of the eigenvalue can be improved by increasing the number of grid points. Stability problems are encountered with the Numerov-Cooley method in the classically forbidden regions [15] and therefore with multi-minima potentials. These problems can be avoided by: ensuring rounding errors are minimised; using global analytical functions to represent the internuclear potential function (thereby avoiding interpolation errors); allowing the domain to be sufficiently large so that the eigenfunctions are not abnormally constrained at the boundaries and using an adequately fine grid. The Numerov-Cooley algorithm has been successfully applied to both single and multi-minima potentials [7-8, 12, 14].

Various other approaches have been used for the solution of the vibrational Schrödinger equation. Traditionally, mathematicians and engineers have used numerical techniques such as the finite-difference and finite-element methods for the solution of Sturm-Liouville type eigenvalue problems. More recently, this approach has been applied to the solution of the vibrational Schrödinger equation [16-19].

In the finite-difference method, the Schrödinger equation is transformed into a set of simultaneous algebraic equations by replacing the derivatives by finite-difference expressions. The accuracys of the calculated eigenvalues are $O(h^2n^4)$ [20]. However, modifications to this method have enabled the accuracy to be improved by reducing the dependence on n [21-22]. The finite-difference method has difficulty coping with irregularities in the potential and therefore has been found to be a less accurate method for solution of the one-dimensional Schrödinger equation than the Numerov-Cooley method [2, 14]. However, the finite-difference method is easily extended to multi-dimensional Schrödinger problems, whereas the Numerov-Cooley method is not as amenable. In order to use the Numerov-Cooley method to study multi-dimensional problems, the system must be considered as a sum of one-dimensional problems (which are each solved using the Numerov-Cooley method) and a perturbation term. The overall problem is solved variationally using the global basis functions obtained from the one-dimensional eigenfunctions. Although the finite-difference method can be used in this manner, it may also be used to directly solve the multi-dimensional problem.

§.16. Accuracy of Finite-Element Method.

The finite-element method (FEM) was developed rapidly in the 1950s and 1960s in order to solve differential equations arising in engineering problems. The method has little difficulty in adapting to complex shapes and irregularities found in engineering problems [23-24]. Moreover, its formulation is based on the calculus of the Rayleigh-Ritz method and so is on more rigorous foundation than finite-difference methods.

The accuracy of the FEM in one-dimension is found to be strongly influenced by the type and order of the basis functions used, the size of the finite-elements, the type of finite-element grid and the global domain used in the solution. Birkhoff et al. [25] have shown that FEM eigenvalues obtained with cubic basis functions are correct to $O(h^6n^8)$, where h is the length of the finite-elements that define the domain for the nth eigenvalue. Therefore, by increasing the number of finite-elements in the domain, the convergence of the eigenvalues is enhanced.

The boundary conditions assume that the eigenfunctions decay to zero at each end of the domain. This assumption may be poor if the domain used does not extend well into the classical forbidden region, resulting in convergence problems and erroneous eigenfunctions and eigenvalues. To ensure that a satisfactory domain is used, it is important to verify that the calculated wavefunction has decayed at each end of the domain. Additionally, the suitability of the domain can be ascertained by recalculating the eigenvalue using a larger domain and applying a Richardson extrapolation as a check on the rate of convergence.

The design of the finite-element grid also affects the precision of the eigenvalues and eigenfunctions. By carefully selecting the size of each element on the domain, a similar precision may be obtained to a calculation which uses a much larger number of equally sized finite-elements [26]. Nevertheless, without a priori knowledge of the potential function it is desirable to keep the size of the finite-elements kept constant over the domain.

Malik et al. [2] compared the Numerov-Cooley method, the finite-difference method and FEM using Lagrangian basis functions for the solution of a one-dimensional Morse oscillator potential. Duff et al. [3] compared spectral methods with finite-element methods for two-dimensional problems for various potentials. In each case, the efficiency of the FEM was found to be comparable with the other methods. An excellent description of FEM is found in Strang and Fix [23].

§.17. The Finite-Element Method Solution of One-dimensional Schrödinger Equation.

Unlike finite-difference methods, the FEM is a variational technique, which resembles the spectral method, but involves discretation of the domain into finite sub-domains called "finite-elements". Its formulation is soundly based on the calculus of the Rayleigh-Ritz and Galerkin methods [23-24].

Using FEM the eigenvalue problem is solved, within each element, using localised basis functions, which are zero outside of the element being considered. Increasing the number of elements in the domain and the number of basis functions per element enables a reduction in complexity of interpolating function. Therefore, in contrast to the spectral methods, whose wavefunction is composed of a large number of complex basis functions, the wavefunction in the FEM is a combination of a small number of simple basis functions, whose coefficients are determined for each element. The FEM gives a flexible solution, which can easily adapt to complex potentials.

A one-dimensional vibration Schrödinger equation can be represented as,

$$\left\{ \hat{T}(Q_i) + \hat{U}_w(Q_i) + \hat{V}(Q_i) \right\} \psi_j(Q_i) = \lambda_j \, \psi_j(Q_i) \qquad (17.1)$$

where Q_i, λ_j and $\psi_j(Q_i)$ represent the mass-weighted i^{th} normal coordinate, j^{th} eigenenergy and eigenfunction respectively. The kinetic energy operator, $\hat{T}(Q_i)$ contains derivatives in Q_i.

The eigenvalues and eigenfunctions of this one-dimensional problem are the stationary points and values of the Rayleigh quotient. Choosing that the eigenfunctions ψ_j are approximated by a trial wavefunction Φ, which is a linear combination of N basis functions ϕ_j,

$$\psi_j(Q_i) \approx \Phi_j(Q_i) = \sum_k^N c_{jk}\, \phi_k(Q_i) \tag{17.2}$$

yields a Rayleigh quotient which may have the form,

$$R(\Phi_j) = \frac{\int \Phi_j{}'\Phi_j{}'\, dQ_i + \int \Phi_j \left(\hat{U}_w + \hat{V}\right) \Phi_j\, dQ_i}{\int \Phi_j^2\, dQ_i} \tag{17.3}$$

The Rayleigh-Ritz finite-element methods are distinctive, because of the choice of local basis functions. The local nature of the finite-element basis functions require definitions of boundaries and corresponding meshes. The finite-element basis functions are a piecewise approximation to the true eigenfunction and so each finite-element basis is only non-zero on a few adjacent intervals. By only considering the subset of basis functions defined on the element of the domain of size 'h', the trial wavefunction on this domain can be expressed in piecewise form as,

$$\Phi_j^h = \sum_{k=1}^n c_{jk}^h\, \phi_k^h \tag{17.4}$$

where n is the number of basis functions defined on the element considered and the superscript 'h' indicates that the basis function is non-zero over part of the element, h. The Rayleigh quotient for the element on the domain $Q_i^h = [0, h]$ becomes,

$$R(\Phi_j^h) = \frac{\sum_k^n \sum_l^n c_{jk}^h c_{jl}^h \int_0^h \left(\phi_k^h{}'\phi_l^h{}' + \phi_k^h \left(\hat{U}_w + \hat{V}\right) \phi_l^h \right) dQ_i^h}{\sum_k^n \sum_l^n c_{jk}^h c_{jl}^h \int_0^h \phi_k^h \phi_l^h\, dQ_i^h} \tag{17.5}$$

This can be written in matrix form as,

$$R^h = \frac{(C^h)^T \, A^h \, C^h}{(C^h)^T \, M^h \, C^h} \tag{17.6}$$

where C^h is the matrix containing the expansion parameters defined on the element h,

$$(C^h)^T = (c_{1j}^h, c_{2j}^h, c_{3j}^h, \ldots, c_{nj}^h) \tag{17.7}$$

and j indicates the jth wavefunction. The A^h matrix is the define on the finite-element and is given by,

$$A_{kl}^h = \int_0^h \left(\phi_k^h{}' \, \phi_l^h{}' + \phi_k^h \left(U_w + \hat{V}(Q_i) \phi_l^h \right) \right) \, dQ_i^h \tag{17.8}$$

Here M^h is the local mass matrix given by,

$$M_{kl}^h = \int_0^h \phi_k^h \phi_l^h \, dQ_i^h \tag{17.9}$$

The stationary values of the Rayleigh quotient given by equation (17.6) yield the solutions of the local eigenvalue problem,

$$A^h C^h = \lambda^h M^h C^h \tag{17.10}$$

which gives the nodal values of the finite-element functions. As many of the overlap integrals over the basis functions are zero, both the global matrices which are assembled from the respective localised matrices A and M are sparse and banded. The matrix M is not diagonal since the basis functions are not orthogonal.

The basis functions used in the FEM are piecewise polynomials which are selected to satisfy certain continuity and completeness properties [3, 23-24]. The completeness property imposes the condition that the basis functions are of at least k^{th} order for a functional which contains derivatives of up the kth order (i.e. the kth order derivatives must exist). To ensure that the solution is physically acceptable, the solution and its derivatives to $(k-1)^{th}$ order should also be continuous. The boundary conditions enforce that the wavefunction is zero at each end of the global domain. In this case the functional is the Rayleigh quotient which contains only the first order derivatives and therefore the basis functions must be at least of first order and the trial wavefunction Φ_i (defined in equation (17.4)) must be continuous.

The simplest local functions which satisfy these conditions are the functions which are linear over an element Q_i^h and equal to zero outside the element (a first order Lagrange function). In this case only two basis functions are defined for each element and hence the optimum trial wavefunction gives a poor representation of the exact wavefunction unless a very large number of finite-elements are used. The functions should vanish at one end of the interval and have a value of unity at the other. More flexible sets of basis functions can be obtained by considering higher order functions [3, 26-27].

Other basis functions which are commonly used are the spline functions [1] and the Hermite functions. These functions satisfy the conditions above as well as extra continuity conditions. In the case of the Hermite functions, the trial function and its derivative are continuous [27]. For spline functions, the trial function and its first and second derivatives are continuous.

The Hermite cubic polynomials have a distinct advantage as basis functions, since they are constructed by imposing continuity on both the trial function and its first derivative. The number of parameters to be determined is therefore reduced. Hermite cubics form a better basis set than more general cubics provided the condition of a continuous first derivative is physically acceptable [23]. The four Hermite cubic functions are given by,

$$\phi_1^h = 1 - 3\left(\frac{Q_i^h}{h}\right)^2 + 2\left(\frac{Q_i^h}{h}\right)^3$$

$$\phi_2^h = Q_i^h - 2\left(\frac{(Q_i^h)^2}{h}\right) + \left(\frac{(Q_i^h)^3}{h^2}\right) \tag{17.11}$$

$$\phi_3^h = 3\left(\frac{Q_i^h}{h}\right)^2 - 2\left(\frac{Q_i^h}{h}\right)^3$$

$$\phi_4^h = -\left(\frac{(Q_i^h)^2}{h}\right) + \left(\frac{(Q_i^h)^3}{h^2}\right)$$

Both functions ϕ_1^h and ϕ_3^h have their first derivatives equal to zero at the end points of the interval. Their value is zero at one end point and unity at the other. The functions ϕ_2^h and ϕ_4^h are equal to zero at each end point and have their first derivatives equal to zero at one end point and unity at the other. To ensure that the functions and their derivatives are continuous, the coefficients of adjacent intervals are related by,

$$c_1^h = c_3^{h-1}$$

$$c_2^h = c_4^{h-1}$$

(17.12)

Figure 5.1 shows the four Hermite cubics defined on any interval, emphasising their local description and the relationship of the coefficients of consecutive intervals, which ensures the continuity of the global wavefunction and its derivative.

Using equation (17.11) the trial wavefunction over an element is [23],

$$\Phi_j^h = a_{j0}^h + a_{j1}^h \, Q_i^h + a_{j2}^h \, (Q_i^h)^2 + a_{j3}^h \, (Q_i^h)^3$$

(17.13)

where **a** is written in matrix form as,

$$a_j^h = h \, c_j^h$$

(17.14)

with the matrix **h** given by,

$$h = \begin{bmatrix} 1 & 0 & 0 & 0 \\ 0 & 1 & 0 & 0 \\ -\dfrac{3}{h^2} & -\dfrac{2}{h} & \dfrac{3}{h^2} & -\dfrac{1}{h} \\ \dfrac{2}{h^3} & \dfrac{1}{h^2} & -\dfrac{2}{h^3} & \dfrac{1}{h^2} \end{bmatrix}$$

(17.15)

Hence,

$$\Phi_j^h = q \, a_j^h$$

(17.16)

where,

$$q = [1, Q_j^h, (Q_j^h)^2, (Q_i^h)^3]$$

(17.17)

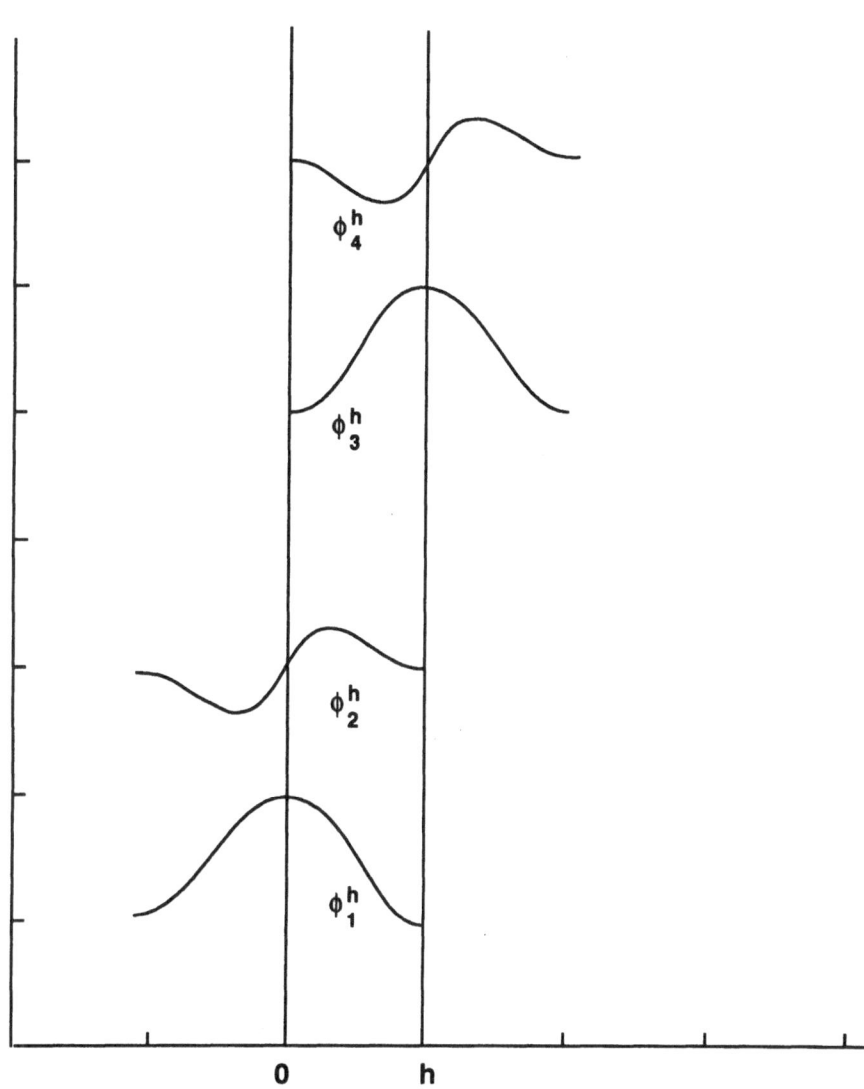

Figure 5.1 The Hermite cubic basis functions.

The numerator of the local Rayleigh quotient becomes,

$$\int_0^h \Phi_j^h{}' \Phi_j^h{}' dQ_i^h + \int_0^h \Phi_j^h \left(\hat{U}_w + \hat{V} \right) \Phi_j^h dQ_i^h$$

$$= \sum_k^4 \sum_l^4 c_{kj}^h c_{lj}^h (K_{kl}^h + P_{kl}^h) \tag{17.18}$$

where K^h is the local stiffness matrix given by,

$$K^h = \frac{1}{30h} \begin{bmatrix} 36 & 3h & -36 & 3h \\ 3h & 4h^2 & -3h & -h^2 \\ -36 & -3h & 36 & -3h \\ 3h & -h^2 & -3h & 4h^2 \end{bmatrix} \tag{17.19}$$

and,

$$P_{kl}^h = \int_0^h \Phi_k^h \left(\hat{U}_w + \hat{V} \right) \Phi_l^h dQ_i^h \tag{17.20}$$

The denominator of the local Rayleigh quotient is,

$$\int_0^h \Phi_j^{h2} dQ_j^h = \sum_k^4 \sum_l^4 c_{ki}^h c_{lj}^h M_{ij}^h \tag{17.21}$$

where the local mass matrix M^h is given by,

$$M^h = \frac{h}{420} \begin{bmatrix} 156 & 22h & 54 & -13h \\ 22h & 4h^2 & 13h & -3h^2 \\ 54 & 13h & 156 & -22h \\ -13h & -3h^2 & -22h & 4h^2 \end{bmatrix} \tag{17.22}$$

The elemental matrices K^h and M^h are evaluated on each interval and are combined to give the global matrices K and M by overlapping the appropriate contributions of one interval with those of the adjacent interval, thereby forming a sparse and banded matrix. The boundary conditions ensure the wavefunctions are zero at the boundaries. That is, c_1^h of the first interval and c_3^h of the last interval are set to zero. The first and last block in the **K** and **M** matrices are therefore of order three, whereas all other blocks are of order four.

The elements of the **P** matrix are evaluated using a Gaussian quadrature scheme. These are added to the global **K** matrix which is then used with the global mass matrix to form the global Rayleigh quotient. Both **A** and **M** matrices are typically large (typically of order 200x200) but they are also sparse. To take advantage of the sparse nature of the matrices a solution of the generalised eigenvalue problem has been developed by Doherty et al. [16] called the incomplete Cholesky conjugate gradient ICCG) method. The **M** matrix is factored into **LL**^{T} by a Cholesky elimination and the eigenvalues of **L**$^{-1}$**A**(**L**$^{-1}$)^T are computed using the QR algorithm [16]. The sequence of conjugate gradient reduction steps reorient the initial random orthogonal vectors to minimise the direction and so hasten convergence. The advantage of the ICCG method is that the sparseness of the **A** and **M** matrices is retained in the **L** matrices because of the use of the iterative factorisation technique [16]. Hence the computer storage capacity required is very much reduced. The Rayleigh quotient is solved using a this diagonalisation routine to give the finite-element eigenfunctions and the eigenvalues.

Various studies on molecular systems have been made using higher order FEMs with multi-dimensional basis functions [3, 26, 28-29]. However, in our variational calculations [16] only the one-dimensional FEM are used, thereby avoiding access to large computing resources.

§.18. The Finite-Square Potential Well.

An example of a one-dimensional motion is a particle of mass m in a square potential well of length a with finite walls of height V_0. It is evident that for $E < V_0$ the spectrum will be discrete, whereas for $E > V_0$ the spectrum is continuous with degenerate energy levels. That is, the potential is defined as,

$$V = 0 \qquad x < 0 ; \xi > a \qquad\qquad (18.1)$$
$$V = -V_0 \qquad 0 \leq \xi \leq a$$

The energy levels of the corresponding one-dimensional Schrödinger equation can be obtained by solving the following equations numerically or graphically [30]:

$$\sin \xi = \pm \gamma \xi \text{ (even states)} \qquad\qquad (18.2)$$
$$\cos \xi = \pm \gamma \xi \text{ (odd states)}$$

where,

$$\xi = a \sqrt{(|V_0|/2)} \text{ and } \gamma = (1/a) \sqrt{(2/|V_0|)} \qquad (18.3)$$

The roots of these two equations determine the energy levels

$$\lambda_n = \xi^2 \frac{2}{ma^2} \qquad (18.4)$$

Of course as $V_0 \to \infty$ the nth eigenenergy and eigenfunction collapse to simpler expressions,

$$\lambda_n = \frac{n^2 \pi^2}{2 m a^2} \qquad (18.5)$$

and

$$\psi_n(x) = \sqrt{\frac{2}{a}} \sin \frac{n \pi x}{a} \qquad (18.6)$$

To demonstrate the usefulness of the FEM in solving problems of this kind, we have chosen a potential for which equation (18.5) is a "good" approximation to the more rigorous equation (18.4) for $n < 3$ and "poorer" for larger n. That is, for a particle of mass 1 au,

$$a = \pi \text{ au}$$

and

$$V_0 = 400 \text{ au}$$

For this simple problem the approximate eigenfunctions are the trigonometric polynomials and so are infinitely accurate at the nodes, although they are in error elsewhere. Generally, it can be shown for this problem that if Hermite cubics are used as interpolating functions in FEM then the eigenfunctions are correct to $O(h^{4-S}n^4)$ in the S derivative and the corresponding leading finite-element eigenvalues are in error to $O(h^6 n^8)$.

Table 5.1 compares the exact eigenenergies with the FEM calculated values using 25, 50 and 100 equivalent finite-elements on the domain $-1.3 \le x \le 4.2$ au. As can be seen from Table 5.1 having h remarkably improves the FEM values. In fact, by using as few as 100 finite-elements, the first and tenth eigenvalues can be determined to within ~0.0001 and 0.01 au of the exact result respectively.

The predicted relative error of FEM dependence on n and h is given by,

$$(\lambda^h - \lambda_n)/ \lambda_n = k h^6 \lambda_n \sim O(h^6 n^6) \qquad (18.7)$$

Table 5.1 Comparison of Analytical and FEM Vibrational Eigenvalues of a Finite-Well Potential.[a]

Energy Level	$E^{(anal)}$ [b] /au	$E^{(FEM)}$ /au [c]		
		25 FE	50 FE	100 FE
1	-399.52177	-399.52288	-399.52062	-399.52170
2	-398.08715	-398.09154	-398.08251	-398.08686
3	-395.69627	-395.70602	-395.68582	-395.69561
4	-392.34941	-392.36639	-392.33075	-392.34823
5	-388.04691	-388.07273	-388.01763	-388.04507
6	-382.78927	-382.82485	-382.74688	-382.78661
7	-376.57708	-376.62207	-376.51902	-376.57345
8	-369.41108	-369.46273	-369.33467	-369.40631
9	-361.29214	-361.34368	-361.19456	-361.28607
10	-352.22133	-352.25949	-352.09949	-352.21377

a) The potential used is defined by equation (18.1) where $a = \pi$ au and $V_0 = 400$ au. Reproduced with permission from reference [17].

b) The analytical eigenvalues for this potential are given by equation (18.5).

c) The numerical eigenvalues were obtained using the FEM with a Hermite cubic basis, using 16 Gaussians in the evaluation of the integrals and a domain of $-1.3 < x < 4.2$ au.

Figure 5.2 gives a logarithmic plot of the relative error as a function of n. The curves are least-squares fits to the discrete data. Clearly the graph follows the predicted functional dependence on n.

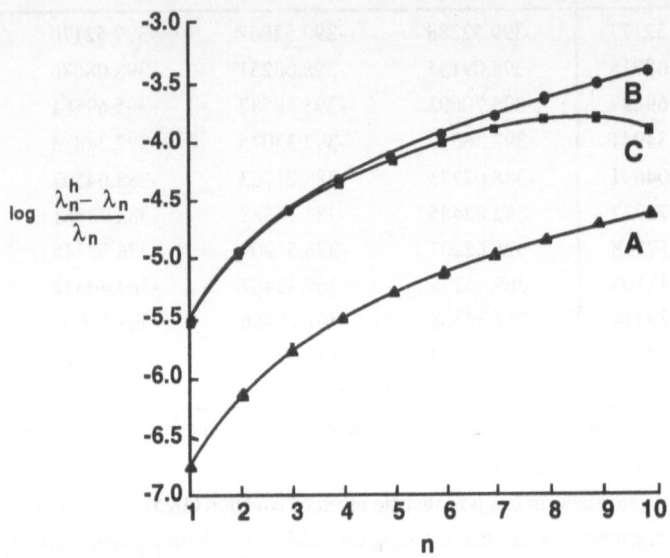

Figure 5.2 Errors in the eigenvalues as a function of their principal quantum number (n). The domain is defined between $-1.3 \leq x \leq 4.2$. The curves are least-squares fits. For each calculation, the lengths of the finite elements are constant. Hence curves A, B and C are the errors associated with 100, 50 and 25 finite elements respectively. Reproduced with permission from reference [17].

The coarse grid (25 finite-elements labelled C) does not parallel the finer grids (labelled A and B) due to a truncation error (i.e. the eigenfunctions do not decay sufficiently in the classical forbidden region). As expected, the coarser grid yields less accurate eigenvalues near the top of the well (i.e. the relative error is translated into a curve that peaks).

Figure 5.3 shows the errors in the eigenfunctions for the first three energy levels (for which equation (18.6) gives a good approximation to the exact eigenfunction). It is clear that by increasing the number of finite-elements, the exact solution can be approached to a desired level of accuracy. The higher energy eigenfunctions for the coarse grid (25 finite-elements) did not decay

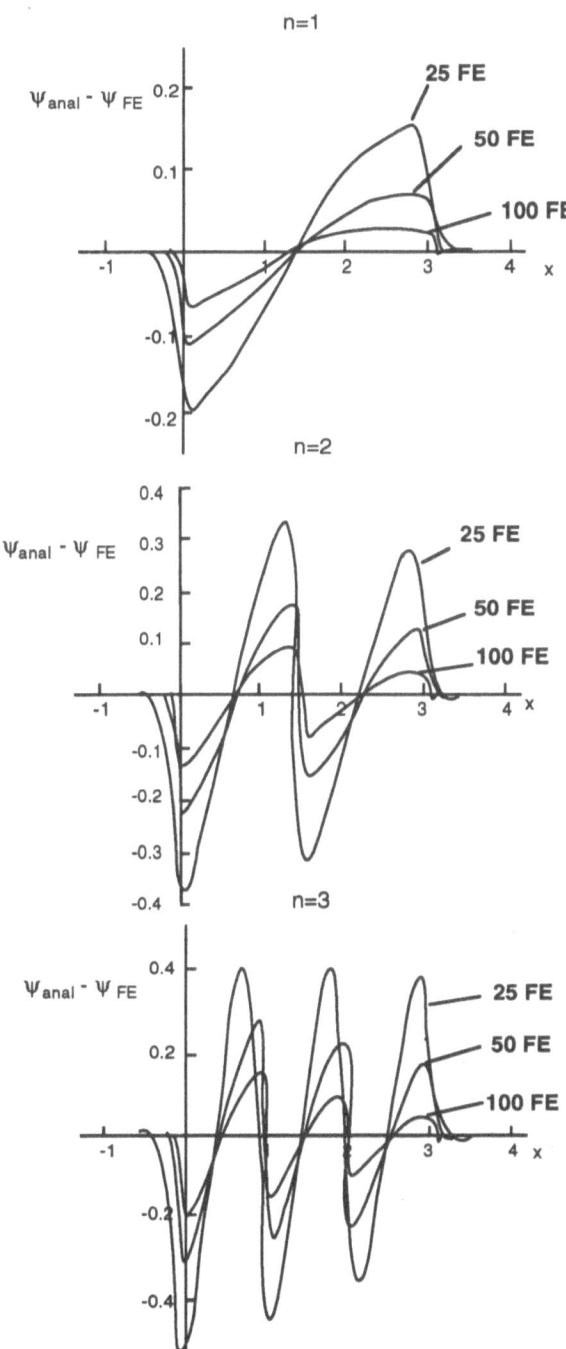

Figure 5.3 Errors in the eigenfunctions for the first three energy levels.

sufficiently at the boundaries of the domain examined, therefore the boundary conditions imposed were not physically realistic resulting in a reduction in accuracy of the solution.

In order to appreciate the convergence properties of FEM it pertinent to realise the more popular Numerov-Cooley method requires over 2000 integration points in order to achieve near single precision accuracy [8]. The slower convergence of Numerov-Cooley method is due to the fact that the eigenvalues are only correct to order $O(h^4)$ rather than $O(h^6)$ as is the case for FEM.

§.19. The Morse Oscillator.

The numerical solution of the Morse potential problem has been discussed by various authors [8, 14, 26-27, 31]. Johnson [8] and Wicke and Harris [14] used the same functional form of the potential,

$$V(x) = D_e \left\{ 1 - \exp\left[-B\left(x - x_e\right)\right] \right\}^2 \tag{19.1}$$

The potential parameters used in this work are those detailed by Wicke and Harris [14]; that is,

$$D_e = 31250 \text{ cm}^{-1}$$

$$B = \left(\frac{8 cm \omega_e \chi_e \pi^2}{h} \right)^{\frac{1}{2}} \times 10^{-8} \text{ Å}^{-1} \tag{19.2}$$

$$\omega_e \chi_e = 8 \text{ cm}^{-1}$$

$$x_e = 1.500 \text{ Å}$$

$$m = 5.00 \text{ g mol}^{-1}$$

The exact Morse eigenenergies for the nth energy level are given by,

$$\lambda_n = 1000\left(n + \frac{1}{2}\right) - 8\left(n + \frac{1}{2}\right)^2 \tag{19.3}$$

and the eigenfunctions constructed from the Hermite cubics are no longer exact at the nodes.

For comparative purposes, the domain chosen was $1.0 \text{ Å} \leq x \leq 2.6 \text{ Å}$ which is the same as that used by Johnson [8]. Table 5.2 gives a comparison of the analytical and the FEM results for the Morse problem. Clearly halving the size of the finite-elements shows a remarkable improve-

Table 5.2 Comparison of Analytical and FEM Vibrational Eigenvalues of a Morse Potential.[a]

Energy Level	$E^{(anal)}$ [b] /cm^{-1}	$E^{(FEM)}$ / cm^{-1} [c]		
		25 FE	50 FE	100 FE
1	498	498.0	498.0	498.0
2	1482	1482.2	1482.1	1482.1
3	2450	2450.5	2450.1	2450.1
4	3402	3403.2	3402.2	3402.2
5	4338	4340.3	4338.3	4338.2
6	5258	5262.0	5258.4	5258.2
7	6162	6170.7	6162.5	6162.3
8	7050	7056.6	7050.7	7050.3
9	7922	7934.6	7922.9	7922.4
10	8778	8812.7	8779.1	8778.4

a) The potential used is defined by equation (19.1). Reproduced with permission from reference [17].

b) The analytical eigenvalues for this potential are given by equation (19.3).

c) The numerical eigenvalues were obtained using the FEM with a Hermite cubic basis, using 16 Gaussians in the evaluation of the integrals and a domain of $-1.0 < x < 2.6$ Å.

ment in the accuracy of the vibrational eigenenergies. Generally, the FEM results are in good agreement with the analytical values with the accuracy decreasing with an increase in the energy level. Using only 100 finite-elements yields an error of less than 0.4 cm^{-1} in the tenth eigenvalue, whereas an error of 34.7 cm^{-1} was obtained using 25 finite-elements. It should be noted that a similar comparison using NCM shows a greater accuracy (error of ~0.00003 cm^{-1}) but at the expense of using 2049 integration points [8].

§.20. Vibrational Analysis and Franck-Condon Factors of Li$_2$.

The use of the FEM to solve stylised Schrödinger problems is a useful means of assessing its ability to calculate vibrational eigenenergies and wavefunctions. However, in order to determine its applicability to the solution of the Schrödinger equation for molecular systems (for which analytical potentials are not available and whose potentials may exhibit localised features), it is preferable to use a real molecular system for which experiment data is available and which have been solved using other numerical solution algorithms.

High quality ab initio potential energy data for the electronic states of Li$_2$ have been reported by Schmidt-Mink et al. [32]. For the ground electronic state ($1^1\Sigma_g^+$) of the lithium dimer their CEPA calculation predicts an internuclear separation of 5.055 au (the experimental value being reported as 5.05 au [33]). Schmidt-Mink et al. [32] also calculated vibrational band origins for each state using the Numerov-Cooley method.

In order to calculate vibrational eigenfunctions and eigenvalues, it is desirable that the discrete ab initio electronic potential is in analytical form. To this end, the discrete potentials of Schmidt-Mink et al.[32] were fitted with cubic splines. Each cubic spline function was plotted to ensure that no irregularities were introduced by the interpolation [3].

Extensive theoretical and experimental data on the low-lying X $1^1\Sigma_g^+$ and A $1^1\Sigma_u^+$ electronic states of Li$_2$ are available in the literature, thereby providing useful comparisons for the results calculated using the FEM with Hermite cubic basis functions. Experimental data is available for various other states, with interesting potential energy curves found for the B $1^1\Pi_u$ (which has a double minimum) and the E $3^1\Sigma_g^+$ state, which has a broad shoulder in the potential. Both these features cause irregular vibrational energy spectra.

Vibrational band origins for the B $1^1\Pi_u$ state and Franck-Condon factors for the X $1^1\Sigma_g^+\leftarrow$B $1^1\Pi_u$ transition, calculated using the FEM with the potential energy surfaces of Schmidt-Mink et al. [32], have been published [34] and were found to be in agreement with experimental results. The first ten vibrational eigenenergies for the X $1^1\Sigma_g^+$ electronic state of Li$_2$ and the first sixteen eigenenergies for the A $1^1\Sigma_u^+$ and E $3^1\Sigma_g^+$ states are presented in Table 5.3. The higher

Table 5.3 Vibrational Eigenenergies for the X $1^1\Sigma_g^+$, A $1^1\Sigma_u^+$ and E $3^1\Sigma_g^+$ Electronic States of Li_2.[a]

n	X $1^1\Sigma_g^+$			A $1^1\Sigma_u^+$				E $3^1\Sigma_g^+$		
	Exp[b]	FEM[c]	[d]	Exp[b]	FEM[c]	[d]	[e]	Exp[f]	FEM[c]	[d]
1	175.0	174.6	175	127.1	127.8	128	126.5	122.1	122.9	123
2	521.3	520.4	521	379.4	380.7	381	377.0	360.1	362.0	362
3	862.3	861.1	861	628.6	630.4	630	624.2	587.5	590.8	591
4	1198.0	1196.5	1197	874.6	876.9	977	868.7	803.2	808.6	808
5	1528.4	1526.5	1527	1117.6	1120.4	1120	1110.5	1007.6	1015.3	1015
6	1853.5	1851.1	1851	1357.4	1360.7	1361	1349.5	1201.6	1211.6	1211
7	2173.1	2170.4	2171	1594.2	1597.8	1598	1585.6	1386.6	1399.0	1399
8	2487.2	2484.2	2484	1828.0	1831.8	1832	1818.6	1563.8	1578.6	1579
9	2795.8	2792.4	2793	2058.6	2062.7	2063	2048.5	1734.3	1751.7	1752
10	3098.7	3094.9	3095	2286.3	2290.6	2291	2275.2	1898.1	1918.1	1918
11				2510.8	2515.3	2515	2498.8	2054.6	2076.6	2077
12				2732.4	2737.0	2737	2719.0	2202.2	2225.0	2225
13				2950.9	2955.7	2956	2936.1	2337.9	2359.4	2359
14				3166.4	3171.2	3171	3150.0		2472.9	2473
15				3378.8	3383.8	3384	3360.8		2553.2	2553
16				3588.1	3593.3	3593	3568.4		2603.5	2603

a) All energies are in cm^{-1}.

b) Experimental eigenenergies obtained from reference [33].

c) Numerical eigenvalues obtained using the FEM and the potential energy surfaces of Schmidt-Mink et al. [32].

d) Numerov-Cooley eigenvalues obtained from reference [32].

e) Numerov-Cooley eigenvalues obtained from reference [35].

f) Experimental eigenenergies obtained from reference [36].

vibrational eigenenergies for the X $1\,^1\Sigma_g^+$ state are not given as the eigenfunctions did not decay at both the boundaries for the domain which was used by Schmidt-Mink et al. [32] and therefore the higher eigenenergies and eigenfunctions calculated for these states would be unreliable. The results obtained by Schmidt-Mink et al. using the Numerov-Cooley method and experimental eigenenergies are also given in Table 5.3. The two sets of calculated eigenenergies are in excellent agreement for the three states, with a maximum deviation of the order of 1 cm^{-1}. The experimental eigenenergies are also in excellent agreement with the calculated results. The maximum deviation of the experimental results from those calculated using the potential of Schmidt-Mink et al. [32] for the A $1\,^1\Sigma_u^+$ state is 5 cm^{-1} for the sixteenth vibrational energy level. The vibrational eigenenergies of the A $1\,^1\Sigma_u^+$ state calculated by Konowalow and Olson [35] using an alternative potential energy surface are also presented. The deviation of the sixteenth eigenenergy from the experimental value is approximately 20 cm^{-1} indicating that the error in the potential energy surface is much greater than that in the solution of the vibrational Schrödinger problem.

In the FEM calculations given above, 2000 finite-elements are used in the domain 3.25 to 30.0 au. The eigenenergies for the ground state were also determined using 4000 finite-elements in order to assess the suitability of the 2000 element grid. It was found that the increase in the number of grid points reduced the tenth eigenenergy from -0.0244748537600 au to -0.0244748537638 au, indicating that the grid was sufficiently fine to obtain accurate eigenenergies. All the vibrational band origins calculated using the FEM [17, 34] are in excellent agreement with those calculated by Schmidt-Mink et al. [32].

Table 5.4 gives the Franck-Condon factors for the X $1\,^1\Sigma_g^+ \leftarrow$ A $1\,^1\Sigma_u^+$ transition of Li$_2$ calculated using the FEM wavefunctions. The Franck-Condon factors are the vibrational overlap integrals given by,

$$FCF(v',v'') = |\int \Psi_{v'} \Psi_{v''} \, d\tau \,|^2 \qquad (20.1)$$

where $\Psi_{v'}$ is the wavefunction for the initial vibrational state and $\Psi_{v''}$ is that for the final state. They are proportional to the band intensity due to the transition. The values in parentheses are those determined experimentally by Kusch and Hessel [36]. Excellent agreement is found between the calculated and experimental results.

The deviation of the calculated from the experimental Franck-Condon factors does not vary greatly with v'. The square root of the sum of the squares of the residuals for transitions from v'=0 is 7×10^{-3} while from v'=15 it is 5×10^{-3}. The variation with v'' is more significant with the square root of the sum of the squares of the residuals for transitions to v''= 0 being 6×10^{-3} while to v''=14 it is 1×10^{-2}. This suggests that the wavefunctions for the ground state decrease in accuracy with an increase in vibrational eigenvalue, whereas the accuracy of those for the A $1\,^1\Sigma_u^+$ state is more uniform. This may be due to the fact that the high energy eigenfunctions of the ground state did not decay in the classical region as well as those of lower energy.

Table 5.5 compares the Franck-Condon factors calculated using the FEM wavefunctions for the A $1\,^1\Sigma_u^+\leftarrow$E $3\,^1\Sigma_g^+$ transition with those determined experimentally [37]. They are found to agree well where experimental data is available. Due to the major departure of the potential from a Morse potential, Bernheim et al. [37] encountered difficulty in the assignment of vibrational levels above $v = 12$. Use of the vibrational band origins calculated by Schmidt-Mink et al. [32] (and confirmed here using the FEM), and the Franck-Condon factors in Table 5.5 should enhance the assignment of higher vibrational energy levels.

Franck-Condon factors for transitions involving all sixteen bound states are given elsewhere [17, 34]. These provide useful aids in the assignment of vibrational levels for those electronic states which are yet to be studied by experimental and other theoretical techniques.

REFERENCES TO CHAPTER V

1 Shore BW (1973) J Chem Phys 58:3855

2 Malik DJ, Eccles J, Secrest D (1980) J Comput Phys 38:157

3 Duff M, Rabitz H, Askar A, Cakmak A, Ablowitz M (1980) J Chem Phys 72:1543

4 Cooley JW (1961) Math Comput 15:363

5 Numerov B (1933) Publ Obs Cent Astrophys Russ 2:188

6 Blatt JM (1967) J Comput Phys 1:382

7 Wolniewicz L, Orlikowski T (1978) J Comput Phys 27:169

8 Johnson BR (1977) J Chem Phys 67:4086

9 Johnson BR (1978) J Chem Phys 69:4678

10 Oset E, Salcedo LL (1985) J Comput Phys 57:361

11 Buedia E, Guardiola R (1985) J Comput Phys 60:561

12 Eckert M (1989) J Comput Phys 82:147

13 Sloan IH (1968) J Comput Phys 2:414

14 Wicke BG, Harris DO (1976) J Chem Phys 64:5236

15 Kolos W,Wolniewicz L (1969) J Chem Phys 50: 3228

16 Doherty G, Hamilton MJ, Burton PG, von Nagy-Felsobuki EI (1986) Aust J Phys 39:749

17 Searles DJ, von Nagy-Felsobuki EI (1988) Am J Phys 56:444

18 Padkjaer SB, Neto JJS, Linderberg J (1992) Chem Phys 161:419

19 Alvarez-Collado JR, Buenker RJ (1992) J Comput Chem 13:135

20 Keller H (1968) Numerical methods for two point boundary value problems, Blaisdell Waltham, New York

21 Anderssen RS, de Hoog FR (1983) Math Scientist 8:115

22 Andrew AL (1989) J Austral Math Soc Ser B 30:460

23 Strang G, Fix GJ (1973) An analysis of the finite element method, Prentice-Hall, New Jersey

24 Norrie DH, de Vries G (1973) The finite element method, Academic Press, New York

25 Birkhoff G, de Boor C, Swartz B, Wendroff B (1966) Siam J Numer Anal 3:188

26 Jaquet R (1990) Comp Phys Commun 58:257

27 Kimura T, Sato N, Suehiro I (1988) J Comput Chem 9:827

28 Sato N, Iwata S (1988) J Comput Chem 9:222

29 Sato N, Iwata S (1988) J Chem Phys 89:2932

30 Eisberg R, Resnick R (1974) Quantum physics of atoms, molecules, solids, nuclei and particles, Wiley, New York

31 Hamilton IP, Light JC (1986) J Chem Phys 84:306

32 Schmidt-Mink I, Müller W, Meyer W (1985) Chem Phys 92:263

33 Hessel MM, Vidal CR (1979) J Chem Phys 70:4439

34 von Nagy-Felsobuki EI, Searles DJ (1991) Franck-Condon factors for 120 transitions involving the lowest-lying 16 vibrational band systems of $^7Li_2$, Australian Institute of Nuclear Science and Engineering, Technical Report, 90/24:1, Sydney

35 Konowalow DD, Olson ML (1979) J Chem Phys 71:450

36 Kusch P, Hessel MM (1977) J Chem Phys 67:586

37 Bernheim RA, Gold LP, Kelly PB, Tipton T, Veirs DK (1982) J Chem Phys 76:57

103

Table 5.4 Comparison of Theoretical and Experimental Franck-Condon Factors for the X $^1\Sigma_g^+ \leftarrow$ A $^1\Sigma_u^+$ Transition of Li$_2$.[a]

v'/v''	0	1	2	3	4	5	6	7
0	.5442D-01(.52D-01)	.1801D+00(.176D+00)	.2728D+00(.270D+00)	.2488D+00(.250D+00)	.1524D+00(.156D+00)	.6573D-01(.68D-01)	.2037D-01(.21D-01)	.4544D-02(.5D-02)
1	.1373D+00(.134D+00)	.1961D+00(.197D+00)	.5415D-01(.58D-01)	.1162D-01(.9D-02)	.1393D+00(.134D+00)	.2117D+00(.211D+00)	.1556D+00(.159D+00)	.6950D-01(.72D-01)
2	.1899D+00(.187D+00)	.7514D-01(.79D-01)	.1774D-01(.15D-01)	.1288D+00(.127D+00)	.5213D-01(.56D-01)	.9915D-02(.8D-02)	.1348D+00(.129D+00)	.1968D+00(.197D+00)
3	.1908D+00(.190D+00)	.2390D-02(.3D-02)	.1004D+00(.98D-01)	.4109D-01(.45D-01)	.2818D-01(.25D-01)	.1119D+00(.112D+00)	.2135D-01(.25D-01)	.3614D-01(.31D-01)
4	.1564D+00(.157D+00)	.2063D-01(.18D-01)	.8798D-01(.90D-01)	.5230D-02(.4D-02)	.9248D-01(.92D-01)	.6522D-02(.9D-02)	.6515D-01(.61D-01)	.8203D-01(.85D-01)
5	.1111D+00(.113D+00)	.7133D-01(.68D-01)	.2450D-01(.27D-01)	.6306D-01(.60D-01)	.2720D-01(.30D-01)	.4244D-01(.39D-01)	.5838D-01(.62D-01)	.7650D-02(.5D-02)
6	.7097D-01(.73D-01)	.1034D+00(.102D+00)	.1613D-03(.000D+00)	.7600D-01(.000D+00)	.3626D-02(.2D-02)	.7033D-01(.71D-01)	.6352D-03(.2D-02)	.7602D-01(.75D-01)
7	.4182D-01(.43D-01)	.1050D+00(.105D+00)	.2348D-01(.21D-01)	.3452D-01(.37D-01)	.4706D-01(.44D-01)	.1628D-01(.19D-01)	.5059D-01(.48D-01)	.1905D-01(.22D-01)
8	.2314D-01(.24D-01)	.8675D-01(.88D-01)	.5878D-01(.56D-01)	.2323D-02(.3D-02)	.6404D-01(.000D+00)	.4122D-02(.3D-02)	.5261D-01(.54D-01)	.9995D-02(.8D-02)
9	.1219D-01(.13D-01)	.6253D-01(.64D-01)	.7916D-01(.78D-01)	.7024D-02(.6D-02)	.3627D-01(.39D-01)	.3922D-01(.37D-01)	.8599D-02(.11D-01)	.5194D-01(.50D-0)
10	.6181D-02(.7D-02)	.4089D-01(.42D-01)	.7947D-01(.80D-01)	.3307D-01(.31D-01)	.6055D-02(.7D-02)	.5476D-01(.55D-01)	.5373D-02(.4D-02)	.3774D-01(.40D-01)
11	.3040D-02(.3D-02)	.2487D-01(.26D-01)	.6673D-01(.68D-01)	.5681D-01(.55D-01)	.1648D-02(.1D-02)	.3477D-01(.37D-01)	.3481D-01(.33D-01)	.3740D-02(.5D-02)
12	.1460D-02(.2D-02)	.1431D-01(.15D-01)	.4962D-01(.51D-01)	.6669D-01(.66D-01)	.1884D-01(.17D-01)	.8494D-02(.10D-01)	.4764D-01(.48D-01)	.6942D-02(.5D-02)
13	.6887D-03(.1D-02)	.7891D-02(.8D-02)	.3377D-01(.35D-01)	.6321D-01(.64D-01)	.4045D-01(.39D-01)	.1686D-03(.000D+00)	.3219D-01(.34D-01)	.3195D-01(.30D-01)
14	.3203D-03(.000D+00)	.4206D-02(.4D-02)	.2150D-01(.22D-01)	.5206D-01(.53D-01)	.5409D-01(.53D-01)	.1090D-01(.9D-02)	.9736D-02(.11D-01)	.4199D-01(.42D-01)
15	.1473D-03(.000D+00)	.2183D-02(.2D-02)	.1300D-01(.14D-01)	.3880D-01(.40D-01)	.5659D-01(.56D-01)	.2895D-01(.27D-01)	.2745D-04(.000D+00)	.2936D-01(.31D-01)

Table 5.4 (Cont.)

v'/v''	8	9	10	11	12	13	14	15
0	.7168D-03(.1D-02)	.7644D-04(.000D+00)	.4999D-05(.000D+00)	.1586D-06(.000D+00)	.0000D+00(.000D+00)	.0000D+00(.000D+00)	.0000D+00(.000D+00)	.0000D+00
1	.2028D-01(.22D-01)	.3911D-02(4D-02)	.4821D-03(.1D-02)	.3442D-04(.000D+00)	.1080D-05(.000D+00)	.0000D+00(.000D+00)	.0000D+00(.000D+00)	.0000D+00
2	.1310D+00(.135D+00)	.5017D-01(.53D-01)	.1177D-01(.13D-01)	.1662D-02(.2D-02)	.1271D-03(.000D+00)	.3712D-05(.000D+00)	.0000D+00(.000D+00)	.0000D+00
3	.1681D+00(.163D+00)	.1784D+00(.180D+00)	.9085D-01(.95D-01)	.2585D-01(.28D-01)	.4140D-02(.5D-02)	.3324D-03(.000D+00)	.8489D-05(.000D+00)	.0000D+00
4	.1462D-05(.000D+00)	.9983D-01(.92D-01)	.1940D+00(.192D+00)	.1343D+00(.139D+00)	.4635D-01(.50D-01)	.8336D-02(.9D-02)	.6881D-03(.1D-02)	.1403D-04
5	.9690D-01(.95D-01)	.2524D-01(.30D-01)	.3602D-01(.30D-01)	.1766D+00(.171D+00)	.1720D+00(.176D+00)	.7199D-01(.77D-01)	.1439D-01(.16D-01)	.1190D-02
6	.6922D-02(.9D-02)	.6145D-01(.57D-01)	.6745D-01(.71D-01)	.3278D-02(.1D-02)	.1373D+00(.128D+00)	.1979D+00(.199D+00)	.1006D+00(.107D+00)	.2208D-01
7	.4303D-01(.39D-01)	.4189D-01(.46D-01)	.1836D-01(.14D-01)	.9001D-01(.91D-01)	.3918D-02(.7D-02)	.9143D-01(.81D-01)	.2104D+00(.208D+00)	.1300D+00
8	.5264D-01(.55D-01)	.7264D-02(.5D-02)	.6650D-01(.67D-01)	.1147D-03(.000D+00)	.8332D-01(.80D-01)	.2514D-01(.31D-01)	.5105D-01(.42D-01)	.2114D+00
9	.1888D-02(.3D-02)	.5672D-01(.55D-01)	.1948D-02(.3D-02)	.6117D-01(.59D-01)	.9969D-02(.13D-01)	.5781D-01(.52D-01)	.5119D-01(.57D-01)	.2248D-01
10	.2206D-01(.19D-01)	.2428D-01(.27D-01)	.3281D-01(.29D-01)	.2178D-01(.25D-01)	.3625D-01(.32D-01)	.3281D-01(.37D-01)	.2930D-01(.24D-01)	.7093D-01
11	.4795D-01(.47D-01)	.1113D-02(.000D+00)	.4517D-01(.47D-01)	.7621D-02(.5D-02)	.4359D-01(.46D-01)	.1175D-01(.9D-02)	.5210D-01(.55D-01)	.8958D-02
12	.2566D-01(.28D-01)	.3090D-01(.29D-01)	.6076D-02(.8D-02)	.4465D-01(.43D-01)	.2071D-03(.1D-02)	.5088D-01(.51D-01)	.4177D-03(.000D+00)	.5929D-01
13	.1102D-02(.2D-02)	.4083D-01(.41D-01)	.8144D-02(.6D-02)	.2505D-01(.28D-01)	.2709D-01(.24D-01)	.1041D-01(.13D-01)	.4208D-01(.39D-01)	.3512D-02
14	.8570D-02(.7D-02)	.1635D-01(.18D-01)	.3485D-01(.33D-01)	.6550D-04(.000D+00)	.3825D-01(.39D-01)	.8343D-02(.6D-02)	.2662D-01(.30D-01)	.2534D-01
15	.2979D-01(.28D-01)	.8021D-04(.000D+00)	.3251D-01(.34D-01)	.1651D-01(.14D-01)	.8941D-02(.11D-01)	.3644D-01(.35D-01)	.1091D-03(.000D+00)	.3757D-01

a) Values in parentheses are obtained from reference [36]. Reproduced with permission from reference [34].

Table 5.5 Franck-Condon Factors for the A $^1\Sigma_u^+ \leftarrow$ E $3^1\Sigma_g^+$ Transition of Li$_2$.[a]

v'/v''	0	1	2	3	4	5	6	7
0	.9965D+00(.996D+00)	.2729D-02(.3D-02)	.7643D-03(.1D-02)	.0000D+00	.5879D-06	.0000D+00	.0000D+00	.0000D+00
1	.2800D-02(.3D-02)	.9938D+00(.994D+00)	.7152D-03(.1D-02)	.2706D-02(.3D-02)	.8391D-05	.4793D-05	.0000D+00	.0000D+00
2	.6894D-03(.1D-02)	.6794D-03(.1D-02)	.9904D+00(.990D+00)	.1131D-02(.1D-02)	.6983D-02(.8D-02)	.1352D-03	.2976D-04	.1365D-05
3	.2238D-05	.2804D-02(.3D-02)	.1900D-02(.2D-02)	.9635D+00(.961D+00)	.1449D-01(.15D-01)	.1622D-01(.18D-01)	.8983D-03(.1D-02)	.1552D-03
4	.2311D-05	.6719D-05	.5987D-02(.6D-02)	.2365D-01(.25D-01)	.8818D+00(.874D+00)	.4911D-01(.50D-01)	.3480D-01(.38D-01)	.3842D-02(.4D-02)
5	.0000D+00	.1451D-04	.2476D-03	.7456D-02(.8D-02)	.8813D-01(.93D-01)	.7227D+00(.709D+00)	.1000D+00(.101D+00)	.6661D-01(.72D-01)
6	.0000D+00	.9284D-06	.2353D-04	.1500D-02(.2D-02)	.3971D-02(.4D-02)	.2035D+00(.213D+00)	.4973D+00(.476D+00)	.1449D+00(.143D+00)
7	.0000D+00	.0000D+00	.9945D-05	.3613D-06	.4334D-02(.5D-02)	.3611D-04	.3424D+00(.355D+00)	.2595D+00(.235D+00)
8	.0000D+00	.0000D+00	.1065D-06	.3405D-04	.1725D-03	.6905D-02(.8D-02)	.1482D-01(.18D-01)	.4434D+00(.451D+00)
9	.0000D+00	.0000D+00	.1252D-06	.4437D-05	.3703D-04	.1370D-02(.2D-02)	.5129D-02(.5D-02)	.7293D-01(.84D-01)
10	.0000D+00	.0000D+00	.0000D+00	.0000D+00	.2625D-04	.1174D-05	.4083D-02(.5D-02)	.1877D-03
11	.0000D+00	.0000D+00	.0000D+00	.1802D-06	.2034D-05	.5111D-04	.3970D-03(.1D-02)	.5845D-02(.7D-02)
12	.0000D+00	.0000D+00	.0000D+00	.0000D+00	.0000D+00	.2107D-04	.5320D-05	.2285D-02(.3D-02)
13	.0000D+00	.0000D+00	.0000D+00	.0000D+00	.2044D-06	.2543D-05	.4121D-04	.2738D-03
14	.0000D+00	.0000D+00	.0000D+00	.0000D+00	.0000D+00	.0000D+00	.2000D-04	.6706D-05
15	.0000D+00	.0000D+00	.0000D+00	.0000D+00	.0000D+00	.0000D+00	.1174D-04	.1920D-05

Table 5.5 (Cont.)

v'/v''	8	9	10	11	12	13	14	15
0	.0000D+00	.0000D+00	.0000D+00	.0000D+00	.0000D+00	.0000D+00	.0000D+00	.0000D+00
1	.0000D+00	.0000D+00	.0000D+00	.0000D+00	.0000D+00	.0000D+00	.0000D+00	.0000D+00
2	.0000D+00	.0000D+00	.0000D+00	.0000D+00	.0000D+00	.0000D+00	.0000D+00	.0000D+00
3	.1149D-04	.8191D-06	.0000D+00	.0000D+00	.0000D+00	.0000D+00	.0000D+00	.0000D+00
4	.6750D-03(.1D-02)	.6985D-04	.6373D-05	.5713D-06	.0000D+00	.0000D+00	.0000D+00	.0000D+00
5	.1204D-01(.13D-01)	.2423D-02(.3D-02)	.3307D-03	.3780D-04	.3864D-05	.3075D-06	.0000D+00	.0000D+00
6	.1106D+00(.119D+00)	.2963D-01(.32D-01)	.7196D-02(.9D-02)	.1247D-02(.2D-02)	.1750D-03	.2133D-04	.2274D-05	.2699D-06
7	.1559D+00(.148D+00)	.1560D+00(.164D+00)	.5931D-01(.65D-01)	.1785D-01(.22D-01)	.3873D-02(.5D-02)	.6726D-03(.1D-02)	.1018D-03	.1435D-04
8	.7986D-01(.62D-01)	.1211D+00(.106D+00)	.1838D+00(.186D+00)	.9888D-01(.107D+00)	.3781D-01(.45D-01)	.1036D-01(.13D-01)	.2265D-02(.3D-02)	.4343D-03(.1D-02)
9	.4431D+00(.435D+00)	.2954D-02	.5780D-01(.41D-01)	.1743D+00(.165D+00)	.1386D+00(.144D+00)	.6986D-01(.80D-01)	.2474D-01(.31D-01)	.6940D-02(.9D-02)
10	.1783D+00(.199D+00)	.3230D+00(.298D+00)	.1944D-01(.28D-01)	.6571D-02(.1D-02)	.1187D+00(.101D+00)	.1576D+00(.154D+00)	.1113D+00(.120D+00)	.5272D-01(.61D-01)
11	.1009D-01(.14D-02)	.2846D+00(.306D+00)	.1388D+00(.119D+00)	.6433D-01(.67D-01)	.1091D-01(.20D-01)	.3810D-01(.27D-01)	.1269D+00(.117D+00)	.1420D+00(.144D+00)
12	.2106D-02(.3D-02)	.6727D-01(.78D-01)	.2917D+00(.332D+00)	.1105D-01(.12D-01)	.5481D-01(.57D-01)	.6533D-01(.73D-01)	.5546D-03	.4327D-01(.52D-01)
13	.4013D-02	.3537D-02	.1599D+00	.1373D+00	.1903D-01	.2351D-02	.6420D-01	.5008D-01
14	.1660D-02	.1826D-03	.4513D-01	.1252D+00	.1583D-02	.2173D-01	.2760D-01	.4120D-05
15	.9141D-03	.1074D-02	.1961D-01	.1118D+00	.1767D-01	.2279D-01	.6802D-02	.1090D-01

a) Values in parentheses are obtained from reference [37]. Reproduced with permission from reference [34].

CHAPTER VI

NUCLEAR SCHRÖDINGER FORMULATION FOR BENT TRIATOMIC SYSTEMS

§.21. Nuclear Schrödinger Formulation.

Development of rovibrational Hamiltonians for even the simple triatomic systems is not routine. As Sutcliffe [1] has pointed out, Eckart's notion of an embedded equilibrium geometry for a stable triatomic molecule necessarily leads to differing rovibrational Hamiltonians for bent [2] and linear [3] nuclear configurations (as derived by Watson). That is, for a bent triatomic molecule with 3N-6 degrees of freedom, singularities in the mass-dependent potential energy operator (coined the "Watson" operator) necessarily cannot yield a smooth transition to a molecule with only 3N-5 degrees of freedom.

Within the normal coordinate representation several approaches were taken to circumvent this problem. Carney et al. [4-6] incorporated the Watson term into their potential-energy expansion by calculating a correction to each grid point of the electronic potential, thereby avoiding singular regions. Bartholomae et al. [7] multiplied their basis functions with auxiliary functions of the form $[1-\exp(-\alpha|R|^2)]$, which vanishes strongly as $|R|$ vanishes. Reimers and Watts [8] employed an auxiliary function of the form $\exp(-10^6|\mu_{\alpha\alpha}^4|)$, where μ is the inertial tensor. Carter and Handy [9] avoided the singularity at linearity by employing Watson's [3] linear Hamiltonian for quasi-linear configurations. Perhaps the most pragmatic approach was that of von Nagy-Felsobuki and coworkers [10-13], who expressed the Watson operator in terms of a Taylor series expansion in normal coordinates (as suggested originally by Watson), thereby developing a D_{3h} model vibrational Hamiltonian [10-11], which is only applicable to small amplitudes of vibration. This model has been extended to the C_{2v} [12] and C_S case [13].

In following section we shall derive the least symmetric rovibrational Hamiltonian for a bent triatomic molecule of C_S symmetry in terms of rectilinear f displacement coordinates (where f is an abbreviation for Japanese word "fukinsei" meaning asymmetry) [13-14]. The Hamiltonian is based on the Eckart-Watson Hamiltonian [2] and by its construction is applicable to only small amplitudes of vibration. Moreover, it is a generalisation of the Carney et al. [4-6] and von Nagy-Felsobuki coworkers [10-12] analysis for D_{3h} and C_{2v} triatomic systems. It is therefore the most general form of the Eckart-Watson Hamiltonian, which is applicable to a bent triatomic system.

§.22. Vibrational Hamiltonian for a C_S Triatomic System in terms of Rectilinear Coordinates.

Figure 6.1 shows the embedded equilibrium geometry of the general bent ABC molecule used in this derivation [14-15]. The right-handed molecule-fixed rectangular coordinate system used has the unit direction vectors $\hat{i}$, $\hat{j}$, $\hat{k}$ orientated so that the $\hat{k}$ direction is orthogonal to the molecular plane, with the coordinate origin at the centre-of-mass of the molecule and with the $\hat{j}$ direction parallel to the base of the triangle (labelled AC in Figure 6.1). The mass and coordinates of atoms A, B and C are labelled by the subscripts 1, 2 and 3 respectively. The equilibrium bond length between atoms A and B is labelled R_1 and R_2 between B and C. The equilibrium angle ABC is $\theta_1 + \theta_2$, with θ_1 and θ_2 defining the angles divided by the normal which passes through atom B to the equilibrium bond AC.

The mass and geometry parameters for the ABC molecule are defined as,

$$M = m_1 + m_2 + m_3 = \text{Total Mass}$$
$$a = \cos \theta_1 \qquad\qquad (22.1)$$
$$b = \sin \theta_1$$
$$c = \sin \theta_2$$

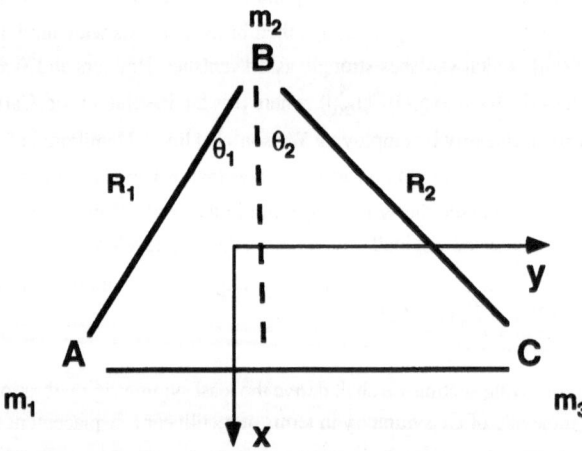

Figure 6.1 The equilibrium geometry of the ABC molecule.

For the ith nucleus, the equilibrium position vectors $\mathbf{r}_i^e$ with components $[\, r_{xi}^e,\ r_{yi}^e,\ r_{zi}^e\,]$ relative to the centre of mass within the Eckart reference frame are derived as,

$$\mathbf{r}_i^e = \begin{bmatrix} r_1^e \\ r_2^e \\ r_3^e \end{bmatrix} = \begin{bmatrix} \dfrac{R_1 a m_2}{M} & \dfrac{m_1 R_1 b - m_3 R_2 c - R_1 b M}{M} & 0 \\[2mm] \dfrac{-R_1 a (m_1 + m_3)}{M} & \dfrac{m_1 R_1 b - m_3 R_2 c}{M} & 0 \\[2mm] \dfrac{R_1 a m_2}{M} & \dfrac{m_1 R_1 b - m_3 R_2 c + R_2 c M}{M} & 0 \end{bmatrix} \begin{bmatrix} \hat{\imath} \\ \hat{\jmath} \\ \hat{k} \end{bmatrix} \qquad (22.2)$$

The instantaneous position vector $\mathbf{r}_i$ in the molecule-fixed system is given by,

$$\mathbf{r}_i = \mathbf{r}_i^e + \Delta \mathbf{r}_i \qquad (22.3)$$

where $\Delta \mathbf{r}_i$ are the Cartesian displacement vectors relative to the equilibrium atomic positions and are given by,

$$\Delta \mathbf{r}_i = \begin{bmatrix} \Delta r_{xi} \\ \Delta r_{yi} \\ \Delta r_{zi} \end{bmatrix} \qquad (22.4)$$

A set of orthonormal rectilinear displacement co-ordinates $\mathbf{f}$ are given as,

$$\mathbf{f} = \begin{bmatrix} f_1 \\ f_2 \\ f_3 \end{bmatrix} \qquad (22.5)$$

and are related to the Cartesian displacement vectors via the transformation,

$$\Delta r_{\alpha i} = \sum_{m=1}^{3} d_{\alpha i m}\, f_m \qquad (22.6)$$

where α is used to represent x, y, and z components and the $d_{\alpha i m}$ are the transformation coefficients.

A more primitive set of rectilinear coordinates, q_k, can be defined by the transformation,

$$\Delta r_{\alpha i} = \sum_{k=1}^{3} A_{\alpha ik} q_k \tag{22.7}$$

The $\mathbf{q}$ and the $\mathbf{f}$ coordinates are related by,

$$q_k = \sum_{m=1}^{3} B_{km} f_m \tag{22.8}$$

and therefore the transformation coefficients are given by,

$$d_{\alpha im} = \sum_{k=1}^{3} A_{\alpha ik} B_{km} \tag{22.9}$$

From the first and second Sayvetz conditions [16] ,

$$\sum_{i=1}^{3} m_i d_{\alpha im} = 0 \tag{22.10}$$

$$\sum_{\beta\gamma} e_{\alpha\beta\gamma} \sum_{i=1}^{3} m_i r_{\beta i}^{e} d_{\gamma im} = 0 \tag{22.11}$$

where $e_{\alpha\beta\gamma}$ are the usual three-dimensional permutation symbols [2]. Hence we can impose the following conditions on the $\mathbf{A}$ transformation coefficients, namely,

$$\sum_{i=1}^{3} m_i A_{\alpha ik} = 0 \tag{22.12}$$

and,

$$\sum_{\beta\gamma} e_{\alpha\beta\gamma} \sum_{i=1}^{3} m_i r_{\beta i}^{e} A_{\gamma ik} = 0 \tag{22.13}$$

Equations (22.12) and (22.13) yield eighteen constraints.

The vibrational or rectilinear coordinates $\mathbf{f}$ are orthonormal and so,

$$\frac{1}{N} \sum_\alpha \sum_{i=1}^3 m_i \, d_{\alpha im} d_{\alpha in} = \delta_{mn} \tag{22.14}$$

The orthonormality condition becomes,

$$\sum_{k=1}^3 \sum_{l=1}^3 \left(\sum_\alpha \sum_{i=1}^3 m_i \, A_{\alpha ik} \, A_{\alpha il} \right) B_{km} \, B_{ln} = N\delta_{mn} \tag{22.15}$$

The condition on the elements of the $\mathbf{B}$ matrix can be simplified by setting,

$$\sum_\alpha \sum_{i=1}^3 m_i A_{\alpha ik} A_{\alpha il} = N'^2 \, \delta_{kl} \tag{22.15}$$

where N' is a constant selected such that the $\mathbf{A}$ matrix is normalised for q_2 and q_3 displacements. That is,

$$\sum_\alpha \sum_{i=1}^3 A_{\alpha i2} \, A_{\alpha i2} = 1 \tag{22.16}$$

Equation (22.15) yields another six constraints on the A coefficients. Hence with twenty-four independent constraints, an additional three constraints need to be imposed in order to define the $\mathbf{A}$ matrix. The three additional constraints are : (a) the displacement of atom A due to the q_2 vibration is directed normal to the equilibrium A-B bond; (b) the displacement of atom C due to the q_2 vibration is directed normal to the equilibrium B-C bond; (c) the displacement of atom A due to the q_3 vibration is directed parallel to the equilibrium A-B bond. Consequently the $\mathbf{A}$ matrix, $[\mathbf{A}_1, \mathbf{A}_2, \mathbf{A}_3]^T$, is derived as,

$$\mathbf{A}_1 = \lambda \begin{bmatrix} -\gamma \dfrac{(R_2 c M_r + R_2 c\alpha + R_1 a\beta)}{m_1} & -\dfrac{R_2 b}{R_1 m_1} & \dfrac{R_2 a}{R_1 m_1} \\[3mm] \gamma \dfrac{(R_1 a M_r - R_1 a\alpha + R_2 c\beta)}{m_1} & -\dfrac{R_2 a}{R_1 m_1} & -\dfrac{R_2 b}{R_1 m_1} \\[3mm] 0 & 0 & 0 \end{bmatrix}$$

$$A_2 = \lambda \begin{bmatrix} \gamma \dfrac{(R_1b + R_2c)\,(M_r + \alpha)}{m_2} & \dfrac{(R_2b + R_1c)}{R_1m_2} & -\dfrac{a(R_2^2 - R_1^2)}{R_1R_2m_2} \\[2ex] -\gamma \dfrac{(R_1b + R_2c)\,\beta}{m_2} & \dfrac{a(R_2^2 - R_1^2)}{R_1R_2m_2} & \dfrac{(R_2b + R_1c)}{R_1m_2} \\[2ex] 0 & 0 & 0 \end{bmatrix} \qquad (22.17)$$

$$A_3 = \lambda \begin{bmatrix} -\gamma \dfrac{(R_1bM_r + R_1b\alpha - R_1a\beta)}{m_3} & -\dfrac{c}{m_3} & -\dfrac{R_1a}{R_2m_3} \\[2ex] -\gamma \dfrac{(R_1aM_r - R_1a\alpha - R_1b\beta)}{m_3} & \dfrac{R_1a}{R_2m_3} & -\dfrac{c}{m_3} \\[2ex] 0 & 0 & 0 \end{bmatrix}$$

where,

$$\lambda = \left\{ \dfrac{1}{R_1^2}\left[\dfrac{R_2^2}{m_1^2} + \dfrac{(R_1b + R_2c)^2}{m_2^2} + \dfrac{R_1^2}{m_3^2} \right] \right\}^{-1/2}$$

$$M_r = \dfrac{\lambda^2}{R_1^2}\left[\dfrac{R_2^2}{m_1} + \dfrac{(R_1b + R_2c)^2}{m_2} + \dfrac{R_1^2}{m_3} \right]$$

$$\alpha = \dfrac{\lambda^2}{R_1^2}\left[\dfrac{R_2^2a^2 - R_2^2c^2}{m_1} - \dfrac{(R_1b + R_2c)^2}{m_2} + \dfrac{R_1^2a^2 - R_1^2b^2}{m_3} \right] \qquad (22.18)$$

$$\beta = \dfrac{2\lambda^2a}{R_1}\left[\dfrac{R_1b}{m_3} - \dfrac{R_2c}{m_1} \right]$$

$$\gamma^2 = \left[R_1^2(M_r^2 - \alpha^2 - \beta^2) \right]^{-1} = \dfrac{m_1m_2m_3}{4a^2\lambda^4M(R_1b + R_2c)^2}$$

Here M_r is the reduced mass for the f rectilinear coordinates and furthermore, N' is equated to $M_r^{1/2}$.

Using the intermediate transformation to the q coordinates, six constraints which can be imposed on the nine elements of the **B** matrix are,

$$\sum_{k=1}^{3} M_r B_{km} B_{kn} = N\delta_{mn} \qquad (22.19)$$

where $m \leq n = 1, 2, 3$.

In order to maximise the diagonal expression for the effective moments of inertia I_{zz}' in terms of only the vibrational coordinates f_1 and the effective vibrational angular momentum $\hat{\pi}_z$ in terms of f_2 and f_3 only the following four constraints can be imposed,

$$B_{12} = B_{13} = 0 \qquad\qquad (22.20)$$

$$B_{31}B_{22} - B_{21}B_{32} = 0 \qquad\qquad (22.21)$$

$$B_{31}B_{23} - B_{21}B_{33} = 0 \qquad\qquad (22.22)$$

The ten conditions given by equations (22.19)-(22.22) are not linearly independent and therefore several solutions can be obtained. The solutions are given by,

$$\mathbf{B} = \begin{bmatrix} \pm(N/M_r)^{1/2} & 0 & 0 \\ 0 & \pm B_{33} & \pm(N/M_r - B_{33}^2)^{1/2} \\ 0 & \pm(N/M_r - B_{33}^2)^{1/2} & B_{33} \end{bmatrix} \qquad (22.23)$$

The simplest form of $\mathbf{B}$ is when $N = M_r$ and $B_{33} = (N/M_r)^{1/2}$ with the signs chosen so that the transformation becomes identical to the D_{3h} cases [4-5] and C_{2v} [6]. Hence the $\mathbf{d}$ matrix, which transforms from the cartesian displacement coordinates to the $\mathbf{f}$ coordinates, is given by $\mathbf{d} = [\mathbf{d}_1, \mathbf{d}_2, \mathbf{d}_3]^T$ where,

$$\mathbf{d}_1 = \lambda \begin{bmatrix} \gamma \dfrac{(R_2 c M_r + R_2 c\alpha + R_1 a\beta)}{m_1} & -\dfrac{R_2 b}{R_1 m_1} & -\dfrac{R_2 a}{R_1 m_1} \\ -\gamma \dfrac{(R_1 a M_r - R_1 a\alpha + R_2 c\beta)}{m_1} & -\dfrac{R_2 a}{R_1 m_1} & \dfrac{R_2 b}{R_1 m_1} \\ 0 & 0 & 0 \end{bmatrix}$$

$$\mathbf{d}_2 = \lambda \begin{bmatrix} -\gamma \dfrac{(R_1 b + R_2 c)(M_r + \alpha)}{m_2} & \dfrac{(R_2 b + R_1 c)}{R_1 m_2} & \dfrac{a(R_2^2 - R_1^2)}{R_1 R_2 m_2} \\ \gamma \dfrac{(R_1 b + R_2 c)\beta}{m_2} & \dfrac{a(R_2^2 - R_1^2)}{R_1 R_2 m_2} & -\dfrac{(R_2 b + R_1 c)}{R_1 m_2} \\ 0 & 0 & 0 \end{bmatrix} \qquad (22.24)$$

$$\mathbf{d}_3 = \lambda \begin{bmatrix} \gamma \dfrac{(R_1 b M_r + R_1 b\alpha - R_1 a\beta)}{m_3} & -\dfrac{c}{m_3} & \dfrac{R_1 a}{R_2 m_3} \\ \gamma \dfrac{(R_1 a M_r - R_1 a\alpha - R_1 b\beta)}{m_3} & \dfrac{R_1 a}{R_2 m_3} & \dfrac{c}{m_3} \\ 0 & 0 & 0 \end{bmatrix}$$

Having defined the coordinate system, the rovibrational Hamiltonian possessing C_S symmetry may be derived. The classical kinetic energy for a triatomic molecule (after the elimination of the centre of mass motion) in terms of the rectilinear f coordinates is,

$$2T = \frac{1}{M_r} \sum_{m=1}^{3} P_m^2 + \sum_{\alpha\beta} (\Pi_\alpha - \pi_\alpha) \, \mu_{\alpha\beta} \, (\Pi_\beta - \pi_\beta) \qquad (22.25)$$

where μ in this case is the reciprocal of the inertial tensor. The total angular momentum is given by,

$$\Pi_\alpha = \frac{\partial T}{\partial \omega_a} = \sum_\beta I_{\alpha\beta} \omega_\beta + \sum_{m=1}^{3} \sum_{n=1}^{3} \zeta_{mn}^{\alpha} f_m \dot{f}_n \qquad (22.26)$$

and the momentum conjugate to the f coordinates given by,

$$P_n = \frac{\partial T}{\partial \dot{f}_n}$$

$$= M_r \dot{f}_n + \sum_\alpha \sum_{m=1}^{3} \zeta_{mn}^{\alpha} \, \omega_\alpha f_m \qquad (22.27)$$

The inertial tensor is defined as,

$$I_{\alpha\beta} = \sum_{\gamma\delta\varepsilon} e_{\alpha\gamma\varepsilon} e_{\beta\delta\varepsilon} \left[\sum_{i=1}^{3} m_i r_{\gamma i}^e r_{\delta i}^e + 2 \sum_{i=1}^{3} m_i r_{\gamma i}^e \sum_{m=1}^{3} d_{\delta im} f_m \right.$$

$$\left. + \sum_{i=1}^{3} m_i \sum_{m=1}^{3} \sum_{n=1}^{3} d_{\gamma im} d_{\delta in} f_m f_n \right] \qquad (22.28)$$

with the Coriolis coupling coefficients given as,

$$\zeta_{mn}^{\alpha} = \sum_{\beta\gamma} e_{\alpha\beta\gamma} \sum_{i=1}^{3} m_i \, d_{\beta im} d_{\gamma in} \qquad (22.29)$$

The vibrational contribution to the total angular momentum, π_α, can be separated from the total angular momentum to give,

$$\Pi_\alpha - \pi_\alpha = \sum_\beta (I_{\alpha\beta} - \sum_{l=1}^{3} \sum_{m=1}^{3} \sum_{n=1}^{3} \zeta_{mn}^{\alpha} \zeta_{ln}^{\beta} \, f_m f_l) \, \omega_\beta$$

$$= \sum_{\beta} I'_{\alpha\beta} \omega_\beta \tag{22.30}$$

where $I'_{\alpha\beta}$ is the effective inertial tensor.

The kinetic energy is now in a form which can be transformed to the quantum mechanical rovibrational Hamiltonian using Podolsky's transformation [17] .The corresponding rovibrational Watson Hamiltonian is given by [3],

$$\hat{H} = \frac{1}{2} \sum_{\alpha\beta} (\hat{\Pi}_\alpha - \hat{\pi}_\alpha) \, \mu_{\alpha\beta} \, (\hat{\Pi}_\beta - \hat{\pi}_\beta) + \frac{1}{2M_r} \sum_{k=1}^{3} \hat{P}_k^2 - \frac{\hbar^2}{8} \sum_{\alpha} \mu_{\alpha\alpha} + \hat{V} \tag{22.31}$$

The third term in this expression is the "Watson" operator.

The rovibrational Hamiltonian can be expressed in terms of the f coordinates yielding,

$$\hat{H} = -\frac{\hbar^2}{2M_r} \sum_{k=1}^{3} \frac{\partial^2}{\partial f_k^2} + \frac{1}{2}\mu_{zz}\hat{\pi}_z\hat{\pi}_z - \mu_{zz}\hat{\pi}_z\hat{\Pi}_z - \frac{\hbar^2}{8} \sum_{\alpha} \mu_{\alpha\alpha} +$$

$$\frac{1}{2}\left\{ \mu_{xx}\hat{\Pi}_x\hat{\Pi}_x + \mu_{yy}\hat{\Pi}_y\hat{\Pi}_y + \mu_{zz}\hat{\Pi}_z\hat{\Pi}_z + \mu_{xy}(\hat{\Pi}_x\hat{\Pi}_y + \hat{\Pi}_y\hat{\Pi}_x)\right\} + \hat{V} \tag{22.32}$$

Zeroing the Coriolis coupling operator, the "purely" vibrational kinetic energy operator is given by,

$$\hat{T}_{vib} = -\frac{\hbar^2}{2M_r} \sum_{k=1}^{3} \frac{\partial^2}{\partial f_k^2} + \frac{1}{2I'_{zz}} \hat{\pi}_z\hat{\pi}_z - \frac{\hbar^2}{8} \sum_{\alpha} \mu_{\alpha\alpha} \tag{22.33}$$

The effective inertial tensor, I'_{zz}, is given by,

$$I'_{zz} = I^e_{zz} + \sum_{m=1}^{3} a^{zz}_m f_m + \frac{1}{4} \sum_{k=1}^{3} \sum_{l=1}^{3} a^{zz}_k (I^{e-1})_{zz} a^{zz}_l f_k f_l \tag{22.34}$$

where,

$$a^{\pi\alpha}_k = 2\sum_{\alpha\gamma} e_{\alpha\gamma\delta} e_{\pi\chi\delta} \sum_{i=1}^{3} m_i \sum_{m=1}^{3} A_{\alpha im} B_{mk} r^e_{\chi i} \tag{22.34}$$

Equation (22.33) yields $a_2^{zz} = a_3^{zz} = 0$ and so effectively diagonalises the effective inertial tensor in terms of f_1 coordinates only. That is,

$$I'_{zz}(f_1) = \frac{m_1 m_2 m_3 M_r R_1^2}{M \lambda^2} + \frac{2 R_1 M_r (m_1 m_2 m_3)^{1/2}}{M^{1/2}} \lambda f_1 + M_r f_1^2 \qquad (22.35)$$

The operator corresponding to the vibrational contribution to the total angular momentum is given by,

$$\hat{\pi}_z = \frac{1}{M_r} \sum_{m=1}^{3} \sum_{n=1}^{3} \zeta_{mn}^z f_m \hat{P}_n$$

$$= -\frac{i\hbar}{M_r} \left\{ \zeta_{12}^z (f_1 \frac{\partial}{\partial f_2} - f_2 \frac{\partial}{\partial f_1}) + \zeta_{13}^z (f_1 \frac{\partial}{\partial f_3} - f_3 \frac{\partial}{\partial f_1}) + \zeta_{23}^z (f_2 \frac{\partial}{\partial f_3} - f_3 \frac{\partial}{\partial f_2}) \right\} \qquad (22.36)$$

From equations (22.21) and (22.22) the Coriolis coefficients ζ_{12}^z and ζ_{13}^z are therefore equal to zero thereby yielding,

$$\hat{\pi}_z \hat{\pi}_z = -\hbar^2 \left(f_3 \frac{\partial}{\partial f_2} - f_2 \frac{\partial}{\partial f_3} \right)^2 \qquad (22.37)$$

The vibrational Hamiltonian is given by,

$$\hat{H}_{vib} = -\frac{\hbar^2}{2M_r} \left\{ \sum_{k=1}^{3} \frac{\partial^2}{\partial f_k^2} + \frac{M_r}{I'_{zz}(f_1)} \left(f_2 \frac{\partial}{\partial f_3} - f_3 \frac{\partial}{\partial f_2} \right)^2 \right\} - \frac{\hbar^2}{8} \sum_{\alpha} \mu_{\alpha\alpha} + \hat{V} \qquad (22.38)$$

where M_r is the reduced mass given above. The first term is the vibrational kinetic energy contribution to the vibrational Hamiltonian, the second term is the vibrational angular momentum contribution and the third term is the Watson term which is a mass dependent contribution to the potential energy operator. The Watson term is the sum of the diagonal elements of the reciprocal effective moment of inertia matrix.

The elements of the μ matrix occur in the Watson and rotational operators. Watson [2] has shown this matrix can be expanded as a Taylor series in terms of the vibrational coordinates. That is,

$$\mu = \mathbf{I}^{e-1} - \mathbf{I}^{e-1} \mathbf{a} \mathbf{I}^{e-1} + \frac{3}{4} \mathbf{I}^{e-1} \mathbf{a} \mathbf{I}^{e-1} \mathbf{a} \mathbf{I}^{e-1} - \frac{1}{2} \mathbf{I}^{e-1} \mathbf{a} \mathbf{I}^{e-1} \mathbf{a} \mathbf{I}^{e-1} \mathbf{a} \mathbf{I}^{e-1} + \dots \qquad (22.39)$$

where,

$$\mathbf{I^{e-1}} = \frac{1}{4(R_1b+R_2c)^2a^2\lambda^2M_r} \begin{bmatrix} 2M_r(M_r+\alpha) & -2M_r\beta & 0 \\ -2M_r\beta & 2M_r(M_r-\alpha) & 0 \\ 0 & 0 & M_r^2-\alpha^2-\beta^2 \end{bmatrix} \qquad (22.40)$$

The elements of the **a** matrix are given by ,

$$\mathbf{a} = \mathbf{a_1}f_1 + \mathbf{a_2}f_2 + \mathbf{a_3}f_3 \qquad (22.41)$$

where,

$$\mathbf{a_1} = \gamma R_1(R_1b+R_2c)2\lambda a \begin{bmatrix} M_r-\alpha & \beta & 0 \\ \beta & M_r+\alpha & 0 \\ 0 & 0 & 2M_r \end{bmatrix}$$

$$\mathbf{a_2} = 2\lambda a \begin{bmatrix} (R_1c+R_2b) & \frac{(R_2^2-R_1^2)a}{R_2} & 0 \\ \frac{(R_2^2-R_1^2)a}{R_2} & -(R_1c+R_2b) & 0 \\ 0 & 0 & 0 \end{bmatrix} \qquad (22.42)$$

$$\mathbf{a_3} = 2\lambda a \begin{bmatrix} \frac{(R_2^2-R_1^2)a}{R_2} & -(R_1c+R_2b) & 0 \\ -(R_1c+R_2b) & -\frac{(R_2^2-R_1^2)a}{R_2} & 0 \\ 0 & 0 & 0 \end{bmatrix}$$

§.23. Vibrational Hamiltonian for a C_{2v} Triatomic System in terms of Rectilinear Coordinates .

Carney et al. [6] has derived the C_{2v} Hamiltonian based on **t** coordinates. The **f** coordinate Hamiltonian given above is a generalisation of this C_{2v} Hamiltonian. The **t** coordinates are therefore rectilinear vibrational coordinates, which have been selected so that the effective moment of inertia and vibrational angular momentum operators are as diagonal as possible.

The Eckart reference frame is embedded within the **t** coordinates. The molecule-fixed coordinate system chosen by Carney et al. [6]. That is, the AB_2 molecule is coplanar with the xy plane; the unit directions of the right-handed coordinate system $\hat{i}$, $\hat{j}$, $\hat{k}$ are chosen with $\hat{i}$ parallel to the C_2 axis and directed away from the A atom and $\hat{k}$ is orthogonal to the xy plane. The mass and coordinates of the central atom A are labelled with a subscript 2, and the mass and coordinates of the B atoms are labelled 1 and 3.

The equilibrium bond length A-B is labelled R and the equilibrium angle BAB is labelled θ. The mass and geometry parameters used in the study of the AB_2 molecules are can be derived from equation (22.1). They are,

$$
\begin{aligned}
R &= R_1 = R_2 \\
\theta &= \theta_1 + \theta_1 \\
M &= m_1 + m_2 + m_3 \\
a &= \cos(\theta/2) \\
b &= c = \sin(\theta/2)
\end{aligned}
\tag{23.1}
$$

From equation (22.18) we can now define,

$$
\lambda = \left(\frac{1}{m_1^2} + \frac{4b^2}{m_2^2} + \frac{1}{m_3^2} \right)^{-1/2}
$$

$$
M_r = \lambda^2 \left(\frac{1}{m_1} + \frac{4b^2}{m_2} + \frac{1}{m_3} \right)
$$

$$
\alpha = \lambda^2 \left(\frac{\cos\theta}{m_1} - \frac{4b^2}{m_2} + \frac{\cos\theta}{m_3} \right)
\tag{23.2}
$$

$$
\beta = \lambda^2 \sin\theta \left(\frac{1}{m_3} - \frac{1}{m_1} \right)
$$

$$
\gamma^2 = \left(1 - \frac{\alpha^2}{M_r^2} - \frac{\beta^2}{M_r^2} \right)
$$

where,

$$
\gamma' = \frac{\lambda}{\gamma M_r R}
\tag{23.3}
$$

The equilibrium atomic position vectors r_i^e relative to the centre-of-mass within the Eckart reference frame from equation (22.2) are given by [6],

$$\mathbf{r}_i^e = \begin{bmatrix} r_1^e \\ r_2^e \\ r_3^e \end{bmatrix} = \frac{R}{M} \begin{bmatrix} am_2 & -b(2m_3+m_2) & 0 \\ -a(m_1+m_3) & b(m_1-m_3) & 0 \\ am_2 & b(2m_1+m_2) & 0 \end{bmatrix} \begin{bmatrix} \hat{i} \\ \hat{j} \\ \hat{k} \end{bmatrix} \qquad (23.4)$$

The **t** coordinates are related to the cartesian coordinates by [6],

$$\mathbf{r}_i = \mathbf{r}_i^e + \mathbf{A}_i \mathbf{B} \mathbf{t} \qquad (23.5)$$

where **B** is given by equation (22.23). From equations (22.23) and (22.17) the $\mathbf{A}_i\mathbf{B}$ matrices (or rather the **d** matrices defined by equation (22.24)) simplify as,

$$\mathbf{d}_1 = \frac{\lambda}{m_1} \begin{bmatrix} \dfrac{b}{\gamma'} + \dfrac{b\alpha}{M_r\gamma'} + \dfrac{a\beta}{M_r\gamma'} & -b & -a \\ \dfrac{-a}{\gamma'} + \dfrac{a\alpha}{M_r\gamma'} - \dfrac{b\beta}{M_r\gamma'} & -a & b \\ 0 & 0 & 0 \end{bmatrix}$$

$$\mathbf{d}_2 = \frac{2\lambda}{m_2} \begin{bmatrix} \dfrac{-b}{\gamma'} - \dfrac{b\alpha}{M_r\gamma'} & b & 0 \\ \dfrac{b\beta}{M_r\gamma'} & 0 & -b \\ 0 & 0 & 0 \end{bmatrix} \qquad (23.6)$$

$$\mathbf{d}_3 = \frac{\lambda}{m_3} \begin{bmatrix} \dfrac{b}{\gamma'} + \dfrac{b\alpha}{M_r\gamma'} - \dfrac{a\beta}{M_r\gamma'} & -b & a \\ \dfrac{-a}{\gamma'} - \dfrac{a\alpha}{M_r\gamma'} - \dfrac{b\beta}{M_r\gamma'} & a & b \\ 0 & 0 & 0 \end{bmatrix}$$

Following the derivation given in §.22. and after separation of the rotational contribution from the quantum mechanical nuclear Hamiltonian, the vibrational Hamiltonian is given by [6],

$$\hat{H}_{vib} = -\frac{\hbar^2}{2M_r} \left\{ \sum_{k=1}^{3} \frac{\partial^2}{\partial t_k^2} + \frac{M_r}{I_{zz}'(t_1)} \left(t_2 \frac{\partial}{\partial t_3} - t_3 \frac{\partial}{\partial t_2} \right)^2 \right\} - \frac{\hbar^2}{8} \sum_{\alpha} \mu_{\alpha\alpha} + \hat{V} \qquad (23.7)$$

where M_r is the reduced mass given by equation (23.2) and where all the operators can now be expressed in terms of the **t** coordinates in an analogous fashion to the **f** coordinate expressions. Here the moment of inertia can be more simply expressed as,

$$I_{zz}'(t_1) = \frac{m_1 m_2 m_3 M_r R^2}{M\lambda^2} + \frac{4R}{\gamma'} \lambda t_1 \sin\theta + M_r t_1^2 \tag{23.8}$$

The C_{2v} equations corresponding to equations (22.40) to (22.43) are,

$$(I^e)^{-1} = \frac{1}{4R^2\lambda^2 M_r \sin^2\theta} \begin{bmatrix} 2M_r(M_r+\alpha) & -2M_r\beta & 0 \\ -2M_r\beta & 2M_r(M_r-\alpha) & 0 \\ 0 & 0 & M_r^2\gamma'^2 \end{bmatrix} \tag{23.9}$$

and the **a** matrix is given by,

$$\mathbf{a} = a_1 t_1 + a_2 t_2 + a_3 t_3 \tag{23.10}$$

where,

$$a_1 = \frac{2R\lambda\sin\theta}{M_r\gamma} \begin{bmatrix} M_r-\alpha & \beta & 0 \\ \beta & M_r+\alpha & 0 \\ 0 & 0 & 2M_r \end{bmatrix}$$

$$a_2 = 2R\lambda\sin\theta \begin{bmatrix} 1 & 0 & 0 \\ 0 & -1 & 0 \\ 0 & 0 & 0 \end{bmatrix} \tag{23.11}$$

$$a_3 = 2R\lambda\sin\theta \begin{bmatrix} 0 & -1 & 0 \\ -1 & 0 & 0 \\ 0 & 0 & 0 \end{bmatrix}$$

§.24. Vibrational Hamiltonian for a D_{3h} Triatomic System in terms of Rectilinear Coordinates.

The D_{3h} Hamiltonian was derived by Carney and Porter [4]. The Hamiltonian can also be extracted from the C_S Hamiltonian, with the proviso that it is realised that the rectilinear coordinates are now also the symmetry coordinates for the system.

The equilibrium bond length A-B is labelled R and the equilibrium angle is labelled θ. The mass and geometry parameters used in the study of the triatomic molecules with D_{3h} symmetry are,

$$
\begin{aligned}
R &= R_1 = R_2 = R_3 \\
\theta &= 60^\circ \\
M &= 3m \\
a &= \cos(\theta/2) = \sqrt{3}/2 \\
b &= c = \sin(\theta/2) = 1/2 \\
\lambda &= m/\sqrt{3} \\
M_r &= m
\end{aligned}
\tag{24.1}
$$

The equilibrium atomic position vectors r_i^e relative to the centre-of-mass within the Eckart reference frame are given by,

$$
r_i^e = \begin{bmatrix} r_1^e \\ r_2^e \\ r_3^e \end{bmatrix} = \frac{R}{3} \begin{bmatrix} a & -3b & 0 \\ -2a & 0 & 0 \\ a & 3b & 0 \end{bmatrix} \begin{bmatrix} \hat{i} \\ \hat{j} \\ \hat{k} \end{bmatrix}
\tag{24.2}
$$

The **s** coordinates are related to the Cartesian coordinates by,

$$
r_i = r_i^e + A_i B s = r_i^e + d_i s
\tag{24.3}
$$

where d_i matrices can be further simplified to,

$$
d_1 = \frac{1}{\sqrt{3}} \begin{bmatrix} b & -b & -a \\ -a & -a & b \\ 0 & 0 & 0 \end{bmatrix}
$$

$$
d_2 = \frac{2}{\sqrt{3}} \begin{bmatrix} -b & b & 0 \\ 0 & 0 & -b \\ 0 & 0 & 0 \end{bmatrix}
\tag{24.4}
$$

$$
d_3 = \frac{1}{\sqrt{3}} \begin{bmatrix} b & -b & a \\ a & a & b \\ 0 & 0 & 0 \end{bmatrix}
$$

After separation of the rotational contribution from the quantum mechanical nuclear Hamiltonian, the vibrational Hamiltonian for D_{3h} is given by [4],

$$\hat{H}_{vib} = -\frac{\hbar^2}{2m}\left\{\sum_{k=1}^{3}\frac{\partial^2}{\partial s_k^2} + \frac{m}{I'_{zz}(s_1)}\left(s_2\frac{\partial}{\partial s_3} - s_3\frac{\partial}{\partial s_2}\right)^2\right\} - \frac{\hbar^2}{8}\sum_\alpha \mu_{\alpha\alpha} + \hat{V} \qquad (24.5)$$

where m is the mass of the nucleus and where,

$$I'_{zz} = m\left\{R^2 + 2Rs_1 + s_1^2\right\} \qquad (24.6)$$

All the operators in equation (24.5) can be expressed in terms of the **s** coordinates, in an analogous fashion to the **t** and **f** coordinate expressions.

The D_{3h} equations analogous to equations (22.40) to (22.42) are,

$$(\mathbf{I}^e)^{-1} = \frac{2}{mR^2}\begin{bmatrix} 1 & 0 & 0 \\ 0 & 1 & 0 \\ 0 & 0 & 1/2 \end{bmatrix} \qquad (24.7)$$

and the **a** matrix is given by,

$$\mathbf{a} = Rm\begin{bmatrix} s_1+s_2 & -s_3 & 0 \\ -s_3 & s_1-s_2 & 0 \\ 0 & 0 & 2s_1 \end{bmatrix} \qquad (24.8)$$

§.25. Rovibrational Hamiltonian for Bent Triatomic Systems.

Equation (22.32) gives the rovibrational Hamiltonian in terms of the **f** coordinates. Setting the Coriolis coupling term to zero, the vibrational Hamiltonian (equation (22.38)) can be used to generate vibrational eigenfunctions which may be utilised in the solution of the rovibrational problem. That is, a set of vibrational eigenfunctions can be calculated which have form,

$$\Psi_m = \sum_{i=1}^{l_1}\sum_{j=1}^{l_2}\sum_{k=1}^{l_3} c_{ijk}^m \psi_i(f_1)\,\psi_j(f_2)\,\psi_k(f_3) \qquad (25.1)$$

where ψ_i is the ith one-dimensional eigenfunction for the **f** coordinate. As a consequence, the rovibrational Hamiltonian may now be expressed in vibration matrix form [5-6],

$$\hat{H}_{mn}^{RV} = E_m <S>_{mn} + 0.5 <A>_{mn} \hat{\Pi}_x^2 + 0.5 <B>_{mn} \hat{\Pi}_y^2 + 0.5 <C>_{mn} \hat{\Pi}_z^2$$
$$+ 0.5 <D>_{mn} \left(\hat{\Pi}_x \hat{\Pi}_y + \hat{\Pi}_y \hat{\Pi}_x \right) + i/h <F>_{mn} \hat{\Pi}_z \qquad (25.2)$$

where E_m is the mth pure vibrational eigenfunction, $<S>_{mn}$ is the overlap vibration matrix element and $\hat{\Pi}$s are the rotational angular-momentum operators, whose components are referred to the molecule-fixed coordinate system. The A-D matrices are symmetric. The <D> and <A> - <B> matrix elements couple adjacent odd or even values of K. The F matrix represents the Coriolis coupling that splits levels with non-zero K values (that is, z axis angular momentum component) in the presence of vibrational angular momentum. The F matrix is given by,

$$<F>_{mn} = <m \mid \frac{\hbar^2}{I_{zz}'} \left(f_3 \frac{\partial}{\partial f_2} - f_2 \frac{\partial}{\partial f_3} \right) \mid n > \qquad (25.3)$$

where I_{zz}' is the inertial tensor.

Similar expression for equations (25.1) to (25.3) can be developed in terms of the s [4] and t [6] coordinate system. Hence our discussion below is applicable to all bent triatomic systems.

In general, the rotational matrix elements are given by [4-6, 10-14],

$$<A>_{mn} = <m \mid \mu_{xx} \mid n > ; <B>_{mn} = <m \mid \mu_{yy} \mid n > ;$$
$$<C>_{mn} = <m \mid \mu_{zz} \mid n > ; <D>_{mn} = <m \mid \mu_{xy} \mid n > \qquad (25.4)$$

where μ is the instantaneous effective reciprocal inertia tensor. The Watson operator is given by the expression [6],

$$\hat{U}_w = -\frac{\hbar^2}{8} (A+B+C) = -\frac{\hbar^2}{8} \sum_\alpha \mu_{\alpha\alpha} \qquad (25.5)$$

Numerical singularities in μ can be circumvented using the perturbation expression given by equation (22.39). For example, in s coordinates the third-order expansion of the Watson term is given by using equation (22.39) in conjunctions with equations (24.7) and (24.8) yielding the expression [10],

$$\hat{U}^{(3)}_w = \frac{\hbar^2}{2m}\left\{\frac{5}{4R^2}\left(-1 + \frac{2s_1}{R} - \frac{3s_1^2}{R^2} + \frac{4s_1^3}{R^3}\right)\right\} + \frac{\hbar^2}{2m}\left\{-\frac{3s_2^2}{R^4} - \frac{3s_3^2}{R^4}\right\} +$$

$$\frac{\hbar^2}{2m}\left\{\frac{12s_1s_3^2}{R^5} + \frac{12s_1s_2^2}{R^5}\right\} \tag{25.6}$$

Similar expression can be derived in t [12, 14] and f [13-14] coordinate systems by using equations (23.9) & (23.11) and equations (22.40) & (22.42) respectively. However, it should not be forgotten that this is a pragmatic approach, since the expansion assumes that μ has no singularities in all regions of the domain.

The vibrationally averaged rotational constants and centrifugal distortions are given by the expectation values of,

$$<A>_{ij} = <i|\mu_{xx}|j> \; ; \; <B>_{ij} = <i|\mu_{yy}|j> \tag{25.7}$$
$$<C>_{ij} = <i|\mu_{zz}|j> \; ; \; <D>_{ij} = <i|\mu_{xy}|j>$$

where i and j label the vibrational eigenfunctions and m is the instantaneous effective reciprocal inertia tensor. Numerical singularities can be circumvented using equation (22.39) to develop a Taylor series expansion for μ. To third-order they are given by,

$$\hat{u}^{(3)}_{zz} = \frac{1}{mR^5}\{R^3 - 2s_1R^2 + 3s_1^2R - 4s_1^3\} \tag{25.8}$$

$$\hat{u}^{(3)}_{xy} = \frac{4}{mR^5}\{R^2s_3 - 3s_1s_3R + 2s_3(3s_1^2 + s_2^2 + s_3^2)\} \tag{25.9}$$

$$\hat{u}^{(3)}_{xx} = \frac{2}{mR^5}\{R^3 - 2R^2(s_1 + s_2) + 3R((s_1 + s_2)^2 + s_3^2) - 4((s_1 + s_2)^3 + (3s_1 + s_2)s_3^2)\} \tag{25.10}$$

$$\hat{u}^{(3)}_{yy} = \frac{2}{mR^5}\{R^3 - 2R^2(s_1 - s_2) + 3R((s_1 - s_2)^2 + s_3^2) - 4((s_1 - s_2)^3 + (3s_1 - s_2)s_3^2)\} \tag{25.11}$$

Similar expression can be derived in t [12, 14] and f [13-14] coordinate systems by using equations (23.9) & (23.11) and equations (22.40) & (22.42) respectively.

The full rovibrational wavefunction Φ is given by linear combination of products of vibrational wavefunctions Ψ_n and symmetric top rotor function ϕ_{jkm},

$$\Phi(\mathbf{v},\mathbf{r})_{jm} = \sum_{n\ k} C_{nkj}\Psi_n(\mathbf{v})\phi_{jkm}(\mathbf{r}) \qquad (25.12)$$

where the $\mathbf{v}$ and $\mathbf{r}$ denote the vibrational and rotational coordinates respectively.

In order to obtain a rovibrational Hamiltonian matrix representation containing real matrix elements, we used the plus and minus combinations of the regular symmetric top eigenfunctions $R_{jkm}^{\pm}$ basis [6, 12],

$$|R_{jkm}^{+}\rangle = 1/\sqrt{2}\ \{\ |\phi_{jkm}(\mathbf{r})\rangle + |\phi_{j\text{-}km}(\mathbf{r})\rangle\}\ \text{and}$$
$$|R_{jkm}^{-}\rangle = 1/\sqrt{2}\ i\ \{\ |\phi_{jkm}(\mathbf{r})\rangle - |\phi_{j\text{-}km}(\mathbf{r})\rangle\} \qquad (25.13)$$

The H^{RV} matrix is constructed using the trial basis functions given by equations (25.7) and diagonalised to yield rovibrational eigenenergies and eigenfunctions. A detailed discussion of the complete solution algorithm is given in Chapter VII.

REFERENCES TO CHAPTER VI

1 Sutcliffe BT (1982) In : Current aspects of quantum chemistry, Carbo R (Ed), Elsevier Scientific Publishing Company, New York
2 Watson JKG (1968) Mol Phys 15:479
3 Watson JKG (1970) Mol Phys 19:465
4 Carney GD, Porter RN (1976) J Chem Phys 65:3547
5 Carney GD, Porter RN (1980) Phys Rev Letts 45:537
6 Carney GD, Langhoff SR, Curtiss LA (1977) J Chem Phys 66:3724
7 Bartholomae R, Martin D, Sutcliffe BT (1981) J Mol Spectrosc 87:367
8 Reimers JR, Watts RO (1984) Mol Phys 52:357
9 Carter S, Handy NC (1982) Mol Phys 47:1445
10 Burton PG, von Nagy-Felsobuki EI, Doherty G, Hamilton M (1984) Chem Phys 83:83
11 Burton PG, von Nagy-Felsobuki EI, Doherty G Chem Phys Lett 104:323
12 Wang F, Searles DJ, von Nagy-Felsobuki EI (1992) J Phys Chem 96:6158
13 Searles DJ, von Nagy-Felsobuki EI (1991) J Chem Phys 95:1107
14 Shaffer WH, Schuman RP (1944) J Chem Phys 12:504
15 Searles DJ (1990) Potential energy surfaces and vibrational band origins of triatomic alkali metal cations, Ph D Thesis, The University of Newcastle, Newcastle
16 Sayvetz A (1939) J Chem Phys 7:383
17 Podolsky B (1928) Phys Rev 32:812

CHAPTER VII

SOLUTION ALGORITHM AND INTEGRAL EVALUATION

§.26. Variational Solution for the Vibrational Schrödinger Equation.

The variational solution algorithm given below is based on the f coordinate Eckart-Watson vibration Hamiltonian (given by equation (22.38)) [1] which collapses to the more symmetric D_{3h} and C_{2v} Hamiltonians. Although the discussion below is written in terms of the f coordinate system, with the appropriate alterations it could equally apply to the solution algorithm of von Nagy-Felsobuki and coworkers [2-3] for the s and t coordinate Hamiltonian.

The Eckart-Watson Hamiltonian in f coordinates may be written as,

$$\hat{H}_{vib} = \sum_{\alpha=1}^{3} \hat{T}_{\alpha} + \hat{T}_{l} + \hat{U}_{w} + \hat{V} \tag{26.1}$$

where $\hat{T}_{\alpha}$ is the kinetic energy operator, $\hat{T}_{l}$ the vibrational angular momentum operator, $\hat{U}_{w}$ the Watson operator and $\hat{V}$ potential energy operator. By restricting the vibration model to small amplitudes of vibration, the Watson term can be expanded as a Taylor series expansion (see §.22.) thereby yielding a model Hamiltonian of form,

$$\hat{H}_{vib} = \sum_{\alpha=1}^{3} \hat{T}_{\alpha} + \hat{T}_{l} + \hat{U}_{w}^{(n)} + \hat{V} \tag{26.2}$$

where n=1, 2, 3..etc. It is evident from equations (22.39) to (22.42) that this model Hamiltonian cannot contain singularities. For the three-dimensional Hamiltonian a third-order expansion for the Watson operator is employed. This ensures convergence of the mass-dependent contribution of the potential energy operator to the vibrational eigenenergies [1-3].

Designing the grid points on the potential energy hypersurface is difficult since only a small grid with sophisticated configuration interaction methods is feasible (see Chapter III). We have employed an adaptive scheme based on the Harris, Engerholm and Gwinn (HEG) [4] potential energy integrator in order to weight the final electronic grid toward the quadrature scheme. Typically surfaces have been constructed using 50-100 ab initio data points [2-3].

An analytical representation of the discrete ab initio electronic potential is not a trivial problem (see Chapter IV). Generally we have employed power series expansions using a variety of expansion variables as well as using Padé approximants [2-3, 5-6].

The three-dimensional vibrational Schrödinger equation is solved variationally using three-dimensional wavefunctions spliced from a set of configuration products of one-dimensional eigenfunctions. That is,

$$\Psi_m = \sum_{i=1}^{n_1} \sum_{j=1}^{n_2} \sum_{k=1}^{n_3} C_{ijk}^m \, \psi_i(f_1) \, \psi_j(f_2) \, \psi_k(f_3) \tag{26.3}$$

where ψ_i is the ith one-dimensional eigenfunction for the a one-dimensional Hamiltonian expressed in a single f coordinate.

The configurational basis functions can be selected using different criteria [7-8]. The nodal selection criterion developed by Carney and Porter [7] ensures that the wavefunction accommodates the most dominant configurations as would be anticipated in the uncoupled mode approximation. That is, the configuration list is determined by imposing the restriction: $i + j + k \leq N$, where N is a nodal cut-off point . Setting N to 13 yields a total of 560 basis functions. Generally, basis set of this size has been employed in a number of calculations [1-3, 8-9]. Although the omission of of higher-order configurations may limit the accuracy of the calculations, 560 basis functions have shown to be reliable with respect to experiment, for example, in variational calculations of the low-lying vibrational states of H_2O^+ [9] and O_3 [10].

The three one-dimensional Hamiltonians are constructed by setting the vibrational angular momentum operator to zero and by using second-order expansion of the Watson operator (which is diagonal in the f coordinates). Only a diagonal representation of the potential energy operator is employed. Hence,

$$\hat{H}(f_1) = -\frac{\hbar^2}{2M_r} \frac{\partial^2}{\partial f_1^2} + \hat{U}_w^{(2)}(f_1) + \hat{V}(f_1)$$

$$\hat{H}(f_2) = -\frac{\hbar^2}{2M_r} \frac{\partial^2}{\partial f_2^2} + \hat{U}_w^{(2)}(f_2) + \hat{V}(f_2) \tag{26.4}$$

$$\hat{H}(f_3) = -\frac{\hbar^2}{2M_r} \frac{\partial^2}{\partial f_3^2} + \hat{U}_w^{(2)}(f_3) + \hat{V}(f_3)$$

The one-dimensional wavefunctions may be numerical (e.g. finite-element or finite-difference methods) or analytical (e.g. Morse and harmonic oscillator functions) or a combination of the two. In our solution algorithm [1-3, 5, 9, 11] we chose the numerical eigenfunctions obtained from finite-element solutions to equation (26.4) using Hermite cubic functions as the local basis set. Usually 1000 finite-elements are employed in each domain. In choosing the integration domain for the three-dimensional problem it is important to ensure that the one-dimensional wavefunctions have decayed correctly in the classical forbidden regions, within this domain.

On applying the variational principle to the secular equation for the three-dimensional vibrational wavefunctions,

$$|H - ES| |C| = 0 \qquad (26.5)$$

we obtain the vibrational eigenvalues E_i and the corresponding configurational expansion coefficients C_{ij}

In order to assess the effective character of the vibrational eigenfunction in terms of the configuration basis, the weight of the basis function in the eigenfunction is used and it is defined as [1-3, 5, 9, 11],

$$(\% \text{ weight})_{ij} = \left\{ \frac{C_{ij}^2}{\sum\limits_{k=1}^{560} C_{ik}^2} \right\}^{\frac{1}{2}} \times 100 \qquad (26.6)$$

where C_{ij} is the jth basis function expansion coefficient in the ith vibrational eigenfunction.

§.27. Evaluation of Integrals.

The evaluation of integrals required in the calculation of vibrational eigenenergies and eigenfunctions depends heavily on the form of he basis set used. For example, Carney and Porter [7] introduced inner projections in order to reduce the powers of the operator in the integrand so to make tractable integrals involving analytical basis functions.

In our solution algorithm [1-3, 5, 9, 11] we use numerical basis functions, which enable all integrals to be calculated without the need to reduce the power of the operators. A detailed description of the numerical schemes used to evaluate integrals will be given below. In order to avoid integrands with long expressions involving the powers of operators, the s coordinate system will be used, although the procedures outlined below have been employed for integral evaluation using both the f [1] and t [3] coordinate systems.

The one-dimensional potential energy integrals can be represented as P_{kl}^h and are calculated numerically using a Gauss quadrature scheme. This scheme is also used for the calculation of kinetic energy, vibrational angular momentum and n^{th}-order Watson integrals involved in solution of the three-dimensional Schrödinger equation. Therefore, a general description of the scheme is given here.

A general form of the integrals evaluated is given by,

$$\int_a^b \phi_k \hat{F}(s_p) \phi_l \, ds_p \tag{27.1}$$

where the la-bl is the magnitude of a finite-element and $\hat{F}(s_p)$ only operates on s_p. The Gauss quadrature scheme approximates the integral by a weighted summation of the integrand at a selected number of quadrature points. Usually a six point quadrature scheme only needs to be employed. The quadrature points and weights are selected so as to maximise the accuracy of the approximation. For the integral given by,

$$\int_0^1 f(y) \, dy \tag{27.2}$$

a suitable selection of the ith six point quadrature points are the zeros of the polynomial [12],

$$q_6(x) = \sqrt{2n + 1} \, P_6^{(k,o)} (1 - 2x) \tag{27.3}$$

where $P_6^{(k,o)}$ are the Jacobi polynomials. The weight factors are given by [12],

$$w_i = \left\{ \sum_{j=0}^5 q_j(x_i)^2 \right\}^{-1} \tag{27.4}$$

In order for the general integral of equation (27.1) to be of the form given by equation (27.2), the integral must first be transformed, giving,

$$\int_a^b \phi_k(s_p) \hat{F}(s_p) \phi_l(s_p) \, ds_p = (b - a) \int_0^1 \phi_k(y) \hat{F}(y) \phi_l(y) \, dy \tag{27.5}$$

where,

$$y = \frac{s_p - a}{b - a} \tag{27.6}$$

Using a six point Gaussian quadrature approximation the integral becomes,

$$\int_0^1 \phi_k(y) \, \hat{F}(y) \, \phi_l(y) \, dy = \sum_{n=1}^{6} w_n \, \phi_k(\rho_n) \, \hat{F}(\rho_n) \, \phi_l(\rho_n) \qquad (27.7)$$

where the ρ_n are the quadrature points which are in the domain [0, 1] and the w_n are the weight factors at each quadrature point [12].

In the case of the integrals P_{ik}^h, $\hat{F}(\rho_k)$ is the sum of the potential energy and the second-order Taylor series expansion of the Watson term at the quadrature point. Therefore, the integrals, P_{ik}^h, can be evaluated by calculating this sum at each quadrature point, summing over the six quadrature points as given by equation (27.7) and multiplying by the transformation constant. They are added to the stiffness matrix for the solution of the one-dimensional Schrödinger equations which determines the coefficients, c_{ik}, in equation (17.2).

The vibrational Hamiltonian given by equation (26.2) is a sum of four operators. Hence, four types of integral need to be evaluated in the solution of the three-dimensional Schrödinger equation. These are the kinetic energy integrals, the potential energy integrals, the vibrational angular momentum integrals and the integrals formed from the third-order expansion of the Watson term.

The three-dimensional potential energy integrals are given by,

$$\langle ijk \,|\, \hat{V} \,|\, lmn \rangle$$

$$= \int\int\int \psi_i(s_1) \, \psi_j(s_2) \, \psi_k(s_3) \, \hat{V}(s_1,s_2,s_3) \, \psi_l(s_1) \psi_m(s_2) \, \psi_n(s_3) \, ds_1 ds_2 ds_3 \qquad (27.8)$$

These integrals can be evaluated using the HEG scheme [4]. In this scheme it is assumed that the electronic potential operator can be expanded as a convergent power series in the coordinates s_1, s_2 and s_3[4]. Hence the potential energy integrals become,

$$\langle ijk \,|\, \hat{V} \,|\, lmn \rangle = \langle ijk \,|\, v_0 \,|\, lmn \rangle + \langle ijk \,|\, v_1 s_1 \,|\, lmn \rangle + \langle ijk \,|\, v_2 s_2 \,|\, lmn \rangle$$

$$+ \langle ijk \,|\, v_3 s_3 \,|\, lmn \rangle + \langle ijk \,|\, v_4 s_1^2 \,|\, lmn \rangle + \langle ijk \,|\, v_5 s_2^2 \,|\, lmn \rangle + \ldots \qquad (27.9)$$

Truncating to first-order the potential energy integrals become,

$$\langle ijk | \hat{V} | lmn \rangle \approx \langle ijk | v_0 | lmn \rangle + \langle ijk \cdot v_1 s_1 | lmn \rangle + \langle ijk | v_2 s_2 | lmn \rangle$$

$$+ \langle ijk | v_3 s_3 | lmn \rangle$$

$$= v_0 \delta_{il} \delta_{jm} \delta_{kn} + v_1 \delta_{jm} \delta_{kn} X_{il}(s_1) + v_2 \delta_{il} \delta_{kn} X_{jm}(s_2) + v_3 \delta_{il} \delta_{jm} X_{kn}(s_3) \quad (27.10)$$

where the elements of the matrix, $X(s_p)$, for each coordinate are given by,

$$X_{ij}(s_p) = \int \psi_i(s_p) s_p \psi_j(s_p) ds_p \quad (27.11)$$

Hence $X(s_p)$ gives the expectation values of the normal coordinate. If $D(s_p)$ is defined as the diagonal form of $X(s_p)$, and $C(s_p)$ are the eigenvectors, then $X(s_p)$ is expressed as [11],

$$X(s_p) = (C(s_p))^T D(s_p) C(s_p) \quad (27.12)$$

The diagonal elements of $X(s_p)$ are the quadrature points of s_p. The expectation values of $X(s_p)$ can be evaluated numerically and therefore the matrix can be diagonalised to determine the quadrature points. The potential energy integrals are determined by evaluating the potential at the quadrature points [11],

$$\langle ijk | \hat{V}(X(s_1), X(s_2), X(s_3)) | lmn \rangle$$

$$= C(s_1) C(s_2) C(s_3) V(D(s_1), D(s_2), D(s_3)) (C(s_1))^T (C(s_2))^T (C(s_3))^T \quad (27.13)$$

The kinetic energy integrals are given by [11],

$$- \frac{\hbar^2}{2m} \langle ijk | \frac{\partial^2}{\partial s_p^2} | lmn \rangle$$

$$= - \frac{\hbar^2}{2m} \int \psi_i(s_q) \psi_l(s_q) ds_q \int \psi_j(s_r) \psi_m(s_r) ds_r \int \psi_k(s_p) \frac{\partial^2}{\partial s_p^2} \psi_n(s_p) ds_p \quad (27.14)$$

for $(p,q,r) = (x,y,z)$ and for the respective permutations.

The vibrational angular momentum integrals in s coordinates are given by [11],

$$- \frac{\hbar^2}{2m} \langle ijk | \frac{1}{(R + s_1)^2} \left\{ s_2 \frac{\partial}{\partial s_3} - s_3 \frac{\partial}{\partial s_2} \right\}^2 | lmn \rangle$$

$$= - \frac{\hbar^2}{2m} \int \psi_i(s_1) \frac{1}{(R + s_1)^2} \psi_l(s_1) ds_1 \left\{ - \int \psi_j(s_2) s_2 \frac{\partial}{\partial s_2} \psi_m(s_2) \, ds_2 \int \psi_k(s_3) \psi_n(s_3) \, ds_3 \right.$$

$$+ \int \psi_j(s_2) s_2^2 \psi_m(s_2) \, ds_2 \int \psi_k(s_3) \frac{\partial^2}{\partial s_3^2} \psi_n(s_3) \, ds_3 + \int \psi_j(s_2) \frac{\partial^2}{\partial s_2^2} \psi_m(s_2) \, ds_2 \int \psi_k(s_3) s_3^2 \psi_n(s_3) \, ds_3$$

$$- \int \psi_j(s_2) s_2 \frac{\partial}{\partial s_2} \psi_m(s_2) \, ds_2 \int \psi_k(s_3) \psi_n(s_3) \, ds_3$$

$$- 2 \int \psi_j(s_2) s_2 \frac{\partial}{\partial s_2} \psi_n(s_2) \, ds_2 \int \psi_k(s_3) s_3 \frac{\partial}{\partial s_3} \psi_m(s_3) \, ds_3 \right\} \qquad (27.15)$$

Four types of one-dimensional integrals are generated by the kinetic energy and vibrational angular momentum operators. They are of the form,

$$\langle i | \hat{G}(s_p) | m \rangle = \int \psi_i \hat{G}(s_p) \psi_m \, ds_p \qquad (27.16)$$

where $\hat{G}(s_p)$ operates on the symmetry coordinate s_p only. The types of one-dimensional integrals generated are: those which contain the second derivative operator; those which contain the first derivative operator; those in which the operator is a function of the symmetry coordinate operators and finally the overlap integrals.

Each of the integrals can be evaluated numerically. Since a finite-element grid is used, the integral given by equation (27.16) can be considered as a sum of the integrals over each finite-element. That is,

$$\int \psi_i \hat{G}(s_p) \psi_m \, ds_p = \sum_{j=1}^{n} \int_{h_{j-1}}^{h_j} \psi_i^j \hat{G}(s_p) \psi_m^j \, ds_p \qquad (27.17)$$

where n is the number of finite-elements and h_j represent the nodal points in the domain. Each of the finite-element wavefunctions can be expressed in terms of the sum of localised basis functions (i.e. in our case the four Hermite cubic basis functions). Therefore, the finite-element integral becomes,

$$\int_{h_{j-1}}^{h_j} \psi_i^j \, \hat{G}(s_p) \, \psi_m^j \, ds_p = \sum_{k=1}^{4} \sum_{l=1}^{4} c_{ik} c_{ml} \int_{h_{j-1}}^{h_j} \phi_k \, \hat{G}(s_p) \, \phi_l \, ds_p \qquad (27.18)$$

This integral is of the form given by equation (27.1). Hence it can be evaluated using Gauss quadrature scheme.

Using equation (27.7), the one-dimensional integrals which contain first derivative operators become,

$$\int \psi_i(s_p) s_p \frac{\partial}{\partial s_p} \psi_m(s_p) \, ds_p$$

$$= \sum_{j=1}^{n} \sum_{k=1}^{4} \sum_{l=1}^{4} c_{ik} c_{ml} \sum_{n=1}^{6} w_n \, \phi_k(\rho_n) \, s_p \frac{\partial}{\partial s_p} \phi_l(\rho_n) \qquad (27.19)$$

Similarly, the one-dimensional integrals containing the second derivative operators are given by,

$$\int \psi_i(s_p) \frac{\partial^2}{\partial s_p^2} \psi_m(s_p) \, ds_p$$

$$= \sum_{j=1}^{n} \sum_{k=1}^{4} \sum_{l=1}^{4} c_{ik} c_{ml} \sum_{n=1}^{6} w_n \, \phi_k(\rho_n) \frac{\partial^2}{\partial s_p^2} \phi_l(\rho_n) \qquad (27.20)$$

Since the first and second derivatives of the Hermite cubics basis functions can be obtained analytically, each of these integrals is readily evaluated.

Using equation (27.7), the one-dimensional integrals in which the operator is a function of the symmetry coordinates is given by,

$$\int \psi_i(s_p) \, f(s_p) \, \psi_m(s_p) \, ds_p$$

$$= \sum_{j=1}^{n} \sum_{k=1}^{4} \sum_{l=1}^{4} c_{ik} c_{ml} \sum_{n=1}^{6} w_n \, \phi_k(\rho_n) \, f(s_p) \, \phi_l(\rho_n) \qquad (27.21)$$

where $f(s_p)$ is a function of s_p.

The overlap integrals are given by,

$$\int \psi_i(s_p)\,\psi_m(s_p)\,ds_p = \sum_{j=1}^{n} \sum_{k=1}^{4} \sum_{l=1}^{4} c_{ik}c_{ml} \sum_{n=1}^{6} w_n\,\phi_k(\rho_n)\,\phi_l(\rho_n) \quad (27.22)$$

In equations (27.19), (27.20) (27.21) and (27.22) the first summation is over all finite-elements, the second and third are over the four Hermite cubic basis functions and the fourth summation is over the six Gauss quadrature points.

The third-order Taylor series expansion of the Watson term is a sum of products of the symmetry coordinate operators and their powers. The integral is given by,

$$\langle\, ijk \,|\, \hat{U}_w^{(3)} \,|\, lmn \,\rangle$$

$$= \frac{\hbar^2}{2m} \int\int\int \psi_i(s_1)\psi_j(s_2)\psi_k(s_3) \left\{ \frac{5}{4R^2}\left(-1 + \frac{2s_1}{R} - \frac{3s_1^2}{R^2} + \frac{4s_1^3}{R^3}\right) - \frac{3s_2^2}{R^4} \right.$$

$$\left. - \frac{3s_3^2}{R^4} + \frac{12s_1s_3^2}{R^5} + \frac{12s_1s_2^2}{R^5} \right\} \psi_l(s_1)\psi_m(s_2)\psi_n(s_3)\,ds_1ds_2ds_3 \quad (27.23)$$

The three-dimensional integral can be expressed as the sum of products of one-dimensional integrals. The one-dimensional integrals are either overlap integrals or contain the position operators. Therefore, they are all of the form given by equation (27.20) and can be evaluated using a Gauss quadrature scheme.

§.28. Analysis of the Vibrational Solution Algorithm.

In the vibrational solution algorithm described in §.26. & §.27., critical decisions need to be made about the number of quadrature points required for the potential energy integrator, the truncation size of the configurational basis functions, the truncation order of the Watson operator and the importance of the various operators given by equation (26.1). All of these factors affect the convergence of the vibrational eigenvalues and eigenfunctions. If the vibrational wavefunctions have not converged, then the rovibrational wavefunctions given by equation (25.12) will be in error. It is therefore important at this point to pause and analyse the vibrational solution algorithm. Such studies were conducted by von Nagy-Felsobuki and coworkers [2, 5, 8, 11] with respect to H_3^+, a molecule which has a $\nu_{0\,1\,\pm1}$ vibrational band origin of 2522 cm^{-1} [13]. As this is the lightest triatomic molecule, such an analysis will expose some of the difficulties associated with the convergence of the vibrational wavefunctions, although for this highly symmetrical molecule there is somewhat restricted mixing of the configurational basis functions due to the different irreducible representations to which they belong. For the more massive triatomics of low symmetry, where the vibrational energy separations are not nearly so large between the different low-lying vibrational levels, configuration mixing is far more severe, making assignments based on the principal configurations fairly tenuous [1].

The choice of the end-points of the domain for the one-dimensional finite-element solutions of requires careful consideration as the eigenfunctions are forced to zero at the end-points. Further, if the end-points are chosen far beyond the region in which the wavefunction has converged then it will not only significantly increase the computation time, but it will produce a three-dimensional domain which goes beyond the realistic behaviour of the physical motion of the molecule. If the end-points chosen are too near, the wavefunction will not have penetrated the classical forbidden region sufficiently and so will not decay to zero, but in fact have a finite amplitude. For H_3^+, Doherty et al. [11] in their finite-element solution used ranges of s_1, s_2 and s_3 of [-1, 2], [-1, 3] and [-1.5, -1.5] respectively. They sub-divided the range into a coarse mesh with constant spacings and then selectively refined the mesh to ensure that the highest eigenvalue required had converged to some tolerance. Table 7.1 gives a mesh for each coordinate and the corresponding eigenvalues of the respective one-dimensional problems. Table 7.1 also gives the Richardson extrapolated results, which further demonstrates that for such a mesh the basis functions have converged. However, in subsequent calculations von Nagy-Felsobuki and coworkers [1-3, 8] for more massive triatomics use a mesh size of 1000 finite-elements of constant size per one-dimensional domain rather than the 28 employed for H_3^+. This was to ensure the one-dimensional wavefunctions accurately reproduced the correct oscillatory behaviour.

The integrals evaluated by gauss quadrature scheme are fairly well posed since a scheme using sixteen quadrature points exactly integrates a polynomial of order seventeen. Hence there is little need to be concerned with numerical inaccuracies in the overlap, kinetic energy and vibrational angular moment integrals, since using a finite-element basis ensures that those integrals involve polynomial of a lesser order. It should be realised that integral involving the $\hat{T}_1$ operator spanned by one-dimensional eigenfunctions reduce to one-dimensional integrals involving the first derivative operator, position operator and dot products thereof. An examination of integrals involving the perturbation expansion of the Watson operator also involves the position operator or dot products thereof. What is however uncertain is the number of quadrature points required by the HEG [4] scheme to effectively integrate the potential energy operator spanned by the vibrational wavefunctions. In this scheme it is difficult to anticipate how many one-dimensional functions in the calculation since they define the maximum number of quadrature points at which the potential integrals are evaluated. Table 7.2 shows the eigenvalues obtained from a 2x2x2 calculation for 5, 10 and 20 quadrature points per dimension obtained for H_3^+. It would appear that the low-lying

Table 7.1 One-Dimensional Eigenvalues (in au) for Normal Modes of H_3^{+a}.

n	s_1^b		s_2^c		s_3^d	
	FEM	Extrap[e]	FEM	Extrap[e]	FEM	Extrap[e]
1	-1.33404	-1.33404	-1.33566	-1.33566	-1.33556	-1.33556
2	-1.31873	-1.31873	-1.32351	-1.32351	-1.32283	-1.32283
3	-1.30387	-1.30387	-1.31186	-1.31186	-1.30991	-1.30991
4	-1.28946	-1.28946	-1.30091	-1.30091	-1.29683	-1.29683
5	-1.27548	-1.27548	-1.29098	-1.29098	-1.28363	-1.28363
6	-1.26192	-1.26192	-1.28260	-1.28260	-1.27034	-1.27034
7	-1.24875	-1.24876	-1.27645	-1.27646	-1.25697	-1.25697
8	-1.23598	-1.23599	-1.27143	-1.27143	-1.24356	-1.24358
9	-1.22358	-1.22360	-1.26600	-1.26608	-1.23013	-1.23017
10	-1.21154	-1.21157	-1.25988	-1.26039	-1.21675	-1.21679

a) Reproduced with permission from reference [11].

b) Interval over range [-1,2] with mesh: 4*0.125, 16*0.0625, 4*0.125, 3*0.25, 0.125.

c) Interval over range [-1,3] with mesh: 3*0.25, 0.125, 20*0.0625, 0.125, 3*0.25.

d) Interval over range [-1.5,1.5] with mesh: 3*0.25, 0.125, 20*0.0625, 0.125, 3*0.125.

e) Richardson extrapolation based on doubling the mesh size to 56 intervals.

Table 7.2 Eigenvalues (in cm^{-1}) from a 2x2x2 Calculation using the HEG quadrature scheme [a].

5x5x5	10x10x10	20x20x20
4418.21	4418.38	4418.38
7003.28	7005.21	7005.22
7005.39	7005.59	7005.59
7899.79	7903.71	7903.74
9782.47	9784.17	9784.18
10760.22	10765.13	10765.17
10816.37	10824.34	10824.41
14441.18	14451.32	14451.44

a) Reproduced with permission from reference [11]. The notation 2x2x2 means that the lowest two one-dimensional eigenfunctions for each coordinate were selected to form the configuration list. The notation nxnxn for the quadrature gives the number of points along each coordinate.

eigenvalues have converged for 10 quadrature points per dimension. However, for high energy vibrational eigenfunctions a 20 point quadrature scheme is used, totally some 8000 quadrature points for the potential energy integrator. The is consistent with the analysis of Cropek and Carney [14].

For H_3^+ the zero-order correction term for the Watson term $\hat{U}_w^{(0)}$ lowers the vibrational energy levels by 27.37 cm^{-1}. Table 7.3 indicates the $\hat{U}_w^{(1)}$, $\hat{U}_w^{(2)}$ and $\hat{U}_w^{(3)}$ corrections lowers the vibrational eigenenergies on an average by 20.87, 27.47 and 27.01 cm^{-1} respectively [2]. Moreover, the average difference $|\hat{U}_w^{(1)}-\hat{U}_w^{(2)}|$ and $|\hat{U}_w^{(2)}-\hat{U}_w^{(3)}|$ is 4.31 and 0.46 cm^{-1} respectively, indicating that for this molecule the $\hat{U}_w^{(3)}$ correction is closely approaching convergence. As the Watson terms are a mass-dependant correction to the potential energy, for the more massive triatomic molecules it would be expected that convergence would be achieve at third-order. Table 7.4 shows the effect of the Watson term truncations (without the $\hat{T}_l$ terms) on the degenerate components of the assigned E states. It is evident that the $\hat{U}_w^{(n)}$ perturbation does not add significantly to the non-degenerancy since it adds ~0.09cm^{-1}.

Table 7.5 shows the effect on the Watson-term truncations (with $\hat{T}_l$ integrals) on the vibrational eigenvalues and eigenfunctions. From the table it is evident that the order of the assignment is independent of the truncations and is completely consistent with Table 7.2. Comparing the %weight of $\hat{H}^{(0)}$ (Table 7.2) and $\hat{H}^{(0)}+\hat{T}_l$ (Table 7.5) indicates that the character of the vibrational wavefunctions has not significantly changed (within 5%) with the inclusion of the $\hat{T}_l$ integrals. However, the inclusion of the $\hat{T}_l$ integrals raises the vibrational energy levels on an average by 27.46 cm^{-1}.

Table 7.5 also indicates that the character of the wavefunction changes only within 0.1% from the first order to the third order corrections. It is also evident that $\hat{U}_w^{(0)}$, $\hat{U}_w^{(2)}$ and $\hat{U}_w^{(3)}$ corrections lowers the vibrational eigenenergies on an average by 23.14, 27.45 and 27.03 cm^{-1} respectively. This is in accordance with the results in Table 7.2. Once again the average difference of $|\hat{U}_w^{(1)}-\hat{U}_w^{(2)}|$ and $|\hat{U}_w^{(2)}-\hat{U}_w^{(3)}|$ is 4.30 and 0.42 cm^{-1} respectively which is in accordance with Table 7.2.

Table 7.6 gives the effect of the Watson term truncations (with $\hat{T}_l$ integrals) on the degenerate components of the E modes. The effect is almost identical to that shown in Table 7.2. The increase of non-degenerancy of the degenerate {(001),(010)}, {($_{(002)}^{200}$), (1011)} and {(101), (110)} components by the $\hat{T}_l$ term is ~0.08, 2.49 and 1.25 cm^{-1} respectively.

The selection criteria used to determine the size of the configuration basis is also of upmost importance. In the case of H_3^+ a number of criteria have been employed [7-8]. Perhaps the simplest

Table 7.3 Effect of Watson Truncation (without the $\hat{T}_1$ integrals) on Vibrational Eigenenergies and Eigenfunctions[a].

Dominant Configuration	$\hat{H}^{(0)}$		$\hat{H}^{(0)} + \hat{U}_w^{(1)}$		$\hat{H}^{(0)} + \hat{U}_w^{(2)}$		$\hat{H}^{(0)} + \hat{U}_w^{(3)}$	
	%W	E_i	%W	E_i	%W	E_i	%W	E_i
000	97	4397.4	97	4372.2	97	4370.4	97	4370.5
001	86	6887.9	86	6863.8	86	6860.5	86	6860.7
010	89	6890.1	89	6866.1	89	6862.7	89	6862.9
100	86	7606.6	87	7583.2	87	7579.7	86	7580.2
(020)+(002)	32+32	9163.6	32+32	9140.5	32+32	9135.3	32+32	9135.6
(020)-(002)	47-33	9316.7	47-33	9293.8	42-33	9288.8	47-33	9289.0
011	63	9324.5	63	9310.6	63	9296.6	63	9296.9
101	64	9955.8	64	9933.6	64	9928.3	64	9928.9
110	66	9981.2	66	9958.9	66	9953.7	66	9954.3
200	77	10710.0	77	10688.4	77	10683.1	77	10684.7

a) Reproduced with permission from reference [2]. The $\hat{H}^{(0)}$ Hamiltonian is defined as : $\Sigma\hat{T}_\alpha + \hat{V}(s_1, s_2, s_3)$, where $\hat{V}$ is a sixth-degree SPF force field . Here the $\hat{U}_w^{(n)}$ designate the different order expansions of the Watson operator. The %W is defined by equation (26.6). All energies are in units of cm^{-1}.

Table 7.4 Effect of Watson-Term Truncation (without the $\hat{T}_1$ integrals) with respect to the Degenerate Components of the E modes [a].

E Modes	$\hat{H}^{(0)}$	$\hat{H}^{(0)} + \hat{U}_w^{(1)}$	$\hat{H}^{(0)} + \hat{U}_w^{(2)}$	$\hat{H}^{(0)} + \hat{U}_w^{(3)}$
(001)-(010)	2.23	2.22	2.22	2.22
$\left\{\begin{matrix}020\\002\end{matrix}\right\}$-(011)	7.75	7.74	7.84	7.84
(101)-(110)	25.38	25.31	25.43	25.41

a) Reproduced with permission from reference [2]. See footnote to table 7.3. The splitting of these components of the E modes is due to the lack of convergence in the 3D vibration expansion. All energies are in units of cm^{-1}.

Table 7.5 Effect of Watson Truncation (with the $\hat{T}_1$ integrals) on Vibrational Eigenenergies and Eigenfunctions[a].

Dominant Configuration	$\hat{H}^{(0)}+\hat{T}_1$		$\hat{H}^{(0)}+\hat{T}_1+\hat{U}_w^{(1)}$		$\hat{H}^{(0)}+\hat{T}_1+\hat{U}_w^{(2)}$		$\hat{H}^{(0)}+\hat{T}_1+\hat{U}_w^{(3)}$	
	%W	E_i	%W	E_i	%W	E_i	%W	E_i
000	97	4397.8	97	4372.6	97	4370.8	97	4370.9
001	86	6910.7	86	6886.7	86	6883.4	86	6883.5
010	89	6913.0	89	6889.0	89	6885.7	89	6885.9
100	86	7607.2	87	7583.8	87	7580.4	87	7580.8
(020)+(002)	33+32	9189.3	33+32	9166.3	33+32	9161.1	33+32	9161.4
(020)-(002)	46-33	9391.4	46-33	9368.6	46-33	9363.5	46-33	9363.8
011	63	9401.7	63	9378.8	63	9373.8	63	9374.1
101	64	9979.7	64	9957.5	64	9952.1	64	9952.8
110	66	10006.3	66	9984.0	66	9978.8	66	9979.5
200	77	10711.1	77.2	10689.6	77	10684.2	77	10685.4

a) Reproduced with permission from reference [2]. See footnote to table 7.3. All energies are in units of cm^{-1}.

Table 7.6 Effect of Watson-Term Truncation (with the $\hat{T}_1$ integrals) with respect to the Degenerate Components of the E modes [a].

E modes	$\hat{H}^{(0)}+\hat{T}_1$	$\hat{H}^{(0)}+\hat{T}_1+\hat{U}_w^{(1)}$	$\hat{H}^{(0)}+\hat{T}_1+\hat{U}_w^{(2)}$	$\hat{H}^{(0)}+\hat{T}_1+\hat{U}_w^{(3)}$
(001)-(010)	2.31	2.29	2.30	2.33
$\left\{\begin{matrix}020\\002\end{matrix}\right\}$-(011)	10.24	10.21	10.33	10.33
(101)-(110)	26.63	26.54	26.68	26.66

a) Reproduced with permission from reference [2]. See footnote to table 7.3 and 7.4. All energies are in units of cm^{-1}.

approach is that employed by Carney and Porter [7]. It is based on a nodal cut-off criterion i.e. $i+j+k$ $\leq N$, where is the cut-off point in terms of a total quantum number and i,j,k label the one-dimensional eigenfunctions used to construct the configurational basis. This leads to $(2+N)!/\{(N-1)!3!$ selected configurations. Table 7.7 compares the N=10 (220 terms) with the N=14 (560 terms) and a full configuration wavefunctions namely the 6x13x8(624-terms) and 6x13x8 (624 terms) for H_3^+ and D_3^+, whereas table 7.8 gives a comparison of the vibrational band origins for these same basis sets [8]. Both tables indicate that the 220 and 560 term wavefunction have in fact converged when compared with the more extensive 624 term wavefunction. For the more massive triatomic molecules von Nagy-Felsobuki and coworkers [1-3, 8] use the N=14 cut-off criterion for their basis set selection and its is evident for D_3^+ that the convergence is more rapid. Carney et al. [9] have previously used this criterion in the selection of a configurational basis for ozone.

§.29. The Rovibrational Solution Algorithm.

The full rovibrational Hamiltonian can be expressed as,

$$\hat{H}^{RV} = \hat{H}_{vib} + \hat{T}_r + \hat{T}_c \qquad (29.1)$$

where $\hat{H}_{vib}$ is defined by equation (26.1), $\hat{T}_r$ is the full rotation kinetic energy operator and $\hat{T}_c$ is the Coriolis operator.

A solution of the rovibrational problem can be achieved if $\Phi(v,r)$ can be expanded in terms of the product vibrational wavefunctions and symmetric top eigenfunctions [9] as,

$$\Phi(v,r)_{jm} = \sum_{n \ k} C_{nkj} \Psi_n(v) \phi_{jkm}(r) \qquad (29.2)$$

where the are expanded in terms of the plus and minus combinations of the regular symmetric top eigenfunctions $R_{jkm}^{\pm}$ basis [9] given by,

$$| R_{jkm}^+ > = 1/\sqrt{2} \{ |\phi_{jkm}(r)> + |\phi_{j-km}(r)> \} \text{ and}$$
$$| R_{jkm}^- > = 1/\sqrt{2} \ i \{ |\phi_{jkm}(r)> - |\phi_{j-km}(r)> \} \qquad (29.3)$$

Expansion (29.2) ensures that rovibrational Hamiltonian matrix will contain real matrix elements and the rovibrational wavefunction is constructed for each value of the rotational angular momentum j and its projection m. Furthermore, for a given j we must sum over k since generally the matrix elements connecting k to k±2 states are non-zero.

Table 7.7 Comparison of Vibrational Eigenenergies as a Function of Configuration Size[a].

Dominant Configuration	$N_{MAX} = 10$ 220-Term	$N_{MAX} = 14$ 560-Term	6x13x8 624-term	10x8x8 640-term
		H_3^+		
000	4363.7	4363.6	4363.6	4363.7
001	6876.8	6875.4	6875.3	6876.4
010	6877.9	6875.7	6875.7	6878.7
100	7573.4	7571.6	7571.6	7573.7
(020)+(002)	9156.0	9139.8	9141.3	9154.2
(020)-(002)	9354.4	9346.5	9348.6	9356.6
011	9373.9	9353.0	9350.3	9367.0
101	9951.6	9940.5	9940.0	9945.7
110	9968.9	9947.8	9946.3	9972.3
200	10679.5	10671.6	10671.3	10678.0
		D_3^+		
000	3111.6	3111.6	3111.6	3111.6
001	4938.4	4938.2	4938.2	4938.3
010	4938.5	4938.2	4938.2	4938.5
100	5437.3	5437.1	5437.1	5437.1
(020)+(002)	6635.4	6632.1	6632.8	6634.6
(020)-(002)	6748.3	6746.9	6747.5	6748.7
011	6751.0	6747.2	6747.2	6749.0
101	7190.1	7187.6	7187.8	7188.5
110	7192.1	7188.5	7188.3	7192.1
200	7706.9	7705.4	7705.4	7706.4

a) Reproduced with permission from reference [8]. All energies are in units of cm[-1].

Table 7.8 Comparison of Vibrational Band Origins as a Function of Configuration Size[a].

Dominant Configuration	$N_{MAX} = 10$ 220-Term	$N_{MAX} = 14$ 560-Term	6x13x8 624-term	10x8x8 640-term
			H_3^+	
$\nu_{0\ 1\ \pm1}$	2514	2512	2512	2514
$\nu_{1\ 0\ 0}$	3210	3208	3208	3210
$\nu_{0\ 2\ 0}$	4792	4776	4778	4791
$\nu_{0\ 2\ \pm2}$	5000	4986	4985	4998
$\nu_{1\ 1\ \pm1}$	5597	5581	5576	5595
$\nu_{2\ 0\ 0}$	6316	6308	6308	6314
			D_3^+	
$\nu_{0\ 1\ \pm1}$	1827	1827	1827	1827
$\nu_{1\ 0\ 0}$	2326	2325	2325	2325
$\nu_{0\ 2\ 0}$	3524	3521	3521	3523
$\nu_{0\ 2\ \pm2}$	3638	3635	3636	3637
$\nu_{1\ 1\ \pm1}$	4080	4076	4076	4079
$\nu_{2\ 0\ 0}$	4595	4594	4594	4595

a) Reproduced with permission from reference [8]. All energies are in units of cm^{-1}.

Carney et al. [10] have detailed the effect of the angular momentum operators acting on the rigid-rotor functions. Nevertheless, for completeness a summary is given here. The projections are,

$$\Pi_z |R_{jkm}^\pm> = \pm ik\hbar \, |R_{jkm}^\mp>$$

$$\Pi_z^2 |R_{jkm}^\pm> = k^2\hbar^2 |R_{jkm}^\pm>$$

$$\Pi_x^2 + \Pi_y^2 + \Pi_z^2 |R_{jkm}^\pm> = j(j+1)\hbar^2 |R_{jkm}^\pm> \qquad (29.4)$$

$$\Pi_x^2 - \Pi_y^2 |R_{jkm}^\pm> = (\hbar^2/2)[(j-k-1)(j-k)(j+k+1)(j+k+2)]^{1/2} |R_{jk+2m}^\pm> +$$

$$(\hbar^2/2)[(j+k-1)(j+k)(j-k+1)(j-k+2)]^{1/2} |R_{jk-2m}^\pm>$$

$$\Pi_x\Pi_y + \Pi_y\Pi_x)|R_{jkm}^\pm> = \overset{-}{\mp} (\hbar^2/2)[(j-k-1)(j-k)(j+k+1)(j+k+2)]^{1/2} |R_{jk+2m}^\mp> \pm$$

$$(\hbar^2/2)[(j+k-1)(j+k)(j-k+1)(j-k+2)]^{1/2} |R_{jk-2m}^\mp>$$

The symmetric top rotor functions are orthogonal. That is,

$$<\phi_{j'k'm'}|\phi_{jkm}> = \delta_{j'j}\delta_{k'k}\delta_{m'm} \qquad (29.5)$$

Equations (29.4) and (29.5) completes the definition of the full rovibrational Hamiltonian in matrix representation. Hence, following Carney et al. [10, 15] the super-matrix to be diagonalised has form,

		First Vib.			Second Vib.		
		k1	k2	...	k1	k2	...
First Vib.	k1						
	k2						
	:						
Second Vib.	k1						
	k2						
	:						

Consequently, there is a matrix of this form for each j and m value since there are not any operators included in the normal coordinate Hamiltonian which will mix states of different total angular momentum. Also since there are no operators that can mix states of different m there is necessarily a (2j+1)-fold degeneracy. Furthermore, since the matrix elements between even and odd k states vanish there are two independent matrices for each j.

The transformation from ϕ to the R basis can be easily carried out using the Wang matrix U_j which has form [16],

$$U_j = 1/\sqrt{2} \begin{bmatrix} -1 & 0 & 0 & 0 & 1 \\ 0 & -1 & 0 & 1 & 0 \\ 0 & 0 & \sqrt{2} & 0 & 0 \\ 0 & 1 & 0 & 1 & 0 \\ 1 & 0 & 0 & 0 & 1 \end{bmatrix} \qquad (29.6)$$

U_j is of order $(2j+1)(2j-1)$ and so be developing H spanned by the transformation of H can easily be achieved using,

$$U_j^{-1} H^{RV} U_j = \tilde{H}^{RV} \qquad (29.7)$$

which thus yields the required factorisation. Diagonalisation yields the eigenvalues and eigenfunctions.

REFERENCE TO CHAPTER VII

1 Searles DJ, von Nagy-Felsobuki EI (1991) J Chem Phys 95:1107
2 Burton PG, von Nagy-Felsobuki EI, Doherty G, Hamilton M (1984) Chem Phys 83:83
3 Wang F, Searles DJ, von Nagy-Felsobuki EI (1992) J Phys Chem 96:6158
4 Harris DO, Engerholm GG, Gwinn WD (1965) J Chem Phys 43:1515
5 Burton PG, von Nagy-Felsobuki EI, Doherty G, Hamilton M (1985) Mol Phys 55:527
6 Searles DJ, von Nagy-Felsobuki EI (1992) Compt Phys Commun 67:527
7 Carney GD, Porter RN (1976) J Chem Phys 65:3547
8 Burton PG, von Nagy-Felsobuki EI, Doherty G (1984) Chem Phys Lett 104:323
9 Wang F, von Nagy-Felsobuki EI (1992) Aust J Phys 45:651
10 Carney GD, Langhoff SR, Curtiss LA (1977) J Chem Phys 66:3724
11 Doherty G, Hamilton M, Burton PG, von Nagy-Felsobuki EI (1986) Aust J Phys 39:749
12 Davis PJ, Polonsky I (1965) In: Handbook of mathematical functions, Abramowitz MA, Stegun IA (Eds), Dover, New York

13 Oka T (1980) Phys Rev Lett 45:531

14 Cropek D, Carney GD (1984) J Chem Phys 80:4280

15 Carney GD, Sprandel LL, Kern CW (1978) Adv Chem Phys 37:305

16 Kroto HW (1975) Molecular rotation spectra, John Wiley & Sons, New York

CHAPTER VIII

DIPOLE MOMENT SURFACES AND RADIATIVE PROPERTIES

§.30. Dipole Moment Surfaces.

There are very few ab initio dipole moment surfaces reported in the literature [1]. It is customary to calculate the partial derivative(s) of the hypersurface surface at small displacements from the equilibrium geometry. To this end it is more convenient to refer the electric dipole moment to the rotating axis system. That is,

$$\mu_S = \sum_\alpha C_{S\alpha} \mu_\alpha \qquad (30.1)$$

where the subscript S refers to the laboratory-fixed X,Y, Z components and the subscript α labels the molecular x, y, z components of the electric dipole moment respectively.

For small amplitudes of vibration, the molecule-fixed components of the dipole moment operator μ_α can be expanded in a power series in the dimensionless normal coordinates q so that,

$$\mu_\alpha = \mu_\alpha^e + \sum_m (\partial\mu_\alpha/\partial q_m)_e q_m + 0.5 \sum_m \sum_n (\partial^2\mu_\alpha/\partial q_m \partial q_n)_e q_m q_n.... \qquad (30.2)$$

where μ_α^e are the components of the permanent dipole moment at the equilibrium geometry. All partial derivatives of the dipole moment with respect to the normal coordinates are referred to the equilibrium geometry of the molecule. The coefficients in the expansion of equation (30.2) are all in Debye (D) units. We would expect that the magnitude of the coefficients to decrease with the expansion order. Typically the magnitude of the coefficients are of order $\mu_\alpha^e \sim 1$, $\partial\mu_\alpha/\partial q_m \sim 0.1$ and $\partial^2\mu_\alpha/\partial q_m^2 \sim 0.01$ [2]. Furthermore, it is clear that the zero-order term is the permanent dipole moment, linear terms being the first-order terms and the electrical anharmonicity of the molecule being contained in the second and higher-order terms.

Producing dipole moment functions from ab initio calculated points may led to non-intuitive coefficients [3]. Further, it is difficult to assess the quality of discrete dipole moment hypersurfaces, since the basis functions, self-consistent field (SCF) procedures and CI methods have been engineered to produce accurate energies, rather than accurate properties such as dipole moments. For example, for some molecules a convergence criterion of 10^{-6} on the square root of the sum of the squares of the elements of the density matrix in an SCF procedure may be sufficient to yield a converged energy value (since this is a slow varying functional), but nevertheless may be still insufficient to produce a converged dipole moment (let alone a reliable one).

In 1974 Green [4] reviewed the errors associated with ab initio calculated dipole moments. He concluded that at the Hartree-Fock limit the error of the dipole moment of diatomic molecules with a single sigma bond is of the order of 0.1 D to 0.2 D. Furthermore, he noted that a similar order of accuracy may be achieved provided that the basis set is at least of a double zeta quality augmented by polarisation functions.

In the case of a molecule like Li_3^+ (in its ground electronic state and at its equilibrium geometry) the centre-of-mass and charge coincide and therefore both $^7Li_3^+$ and $^6Li_3^+$, do not possess a permanent dipole moment. However, for displaced geometries, which do not contain a C_3 axis these molecules have a non-zero permanent dipole moment. Other isotopic variants of this molecule and triatomic molecules of lower symmetry possess a permanent dipole moment, even at their equilibrium geometry, due to the displacement of mass and charge centres.

A 61 point discrete dipole moment surface for Li_3^+ as a function of the three normal coordinates has been reported [5] and shall serve as an example here of the difficulties in fitting such surfaces. The geometry dependent dipole moments were calculated for $^7Li_3^+$ and $^6Li_3^+$ with respect to an origin at the centre of the nuclear charge. The surface was calculated at the Hartree-Fock SCF level using the GAUSSIAN 82 package [5] employing the [11s,3p,1d/6s,3p,1d] basis of Gerber and Schumacher [6] with d orbital exponent being partially optimised at 0.15 [5]. The basis set meets the criterion set by Green [4], although it is deficient with respect to producing a reliable Hartree-Fock limit. Nevertheless, this molecule has sigma bonds and so the error of data points on the dipole hypersurface are likely to be within ±0.2 D.

The grid for the dipole moment surface differed from that used for the electronic energy calculations [5], since the data points for purely S_1 vibrations gave dipole moments of zero in both the x and y direction and so were not included in the surface. Further, in order to more accurately define the surface, extra off-diagonal data points (that is, those due to combinations of S_2 and S_3 coordinates) were calculated.

A major difficulty in producing a dipole moment function of D_{3h} triatomic molecules is the near rank deficiencies involving the S_1 cross-terms in a power series expansion of a discrete ab initio dipole moment surface [3, 5, 7]. Table 8.1 yields a regression analysis in fitting a 61 point discrete ab initio dipole moment surface of Li_3^+ to the dipole moment function expanded as [3],

Table 8.1 Regression Analysis of a Power Series Fit to the Ab Initio Dipole Moment Surface of Li_3^{+a}.

Expansion Parameter		Coefficients		
μ_x	μ_y	I[b]	II[c]	III[d]
S_2	S_3	0.30664	0.30467	0.30664
$0.5(S_3^2 - S_2^2)$	$S_2 S_3$	0.16586	0.17048	0.16586
$S_1 S_2$	$S_1 S_3$	-	8077.86393	0.00000
$S_2(S_2^2 + S_3^2)$	$S_3(S_2^2 + S_3^2)$	0.00223	0.00343	0.00223
$0.5 S_1(S_3^2 - S_2^2)$	$S_1 S_2 S_3$	-	10098.50668	0.00000
$0.125(S_2^2 - S_3^2 + 2S_2 S_3)$ $(S_3^2 - S_2^2 + 2S_2 S_3)$	$S_2 S_3(S_3^2 + S_2^2)$	-0.00110	0.001120	-0.00110
$S_1 S_2(S_2^2 + S_3^2)$	$S_1 S_3(S_2^2 + S_3^2)$	-	294.97575	0.00000
$3.5 S_2^5 + 12 S_2^3 S_3^2 + 2 S_2 S_3^4$	$3.5 S_3^5 + 12 S_3^3 S_2^2 + 2 S_3 S_2^4$	0.00003	0.00002	0.00003
$0.125 S_1(S_2^2 - S_3^2 + 2S_2 S_3)$ $(S_3^2 - S_2^2 + 2S_2 S_3)$	$S_1 S_2 S_3(S_3^2 + S_2^2)$	-	-532.66357	0.00000
	χ^2	0.27305	0.24967	0.27305

a) Reproduced with permission from reference [3]. See reference [5] for a more detailed description of the SCF ab initio dipole moment surfaces.
b) No S_1 cross-terms included in the fit. See reference [5].
c) Including S_1 cross-terms but no SVD.
d) SVD in which all S_1 cross-terms singular values set to zero.

$$\mu = C_1[S_2\hat{i} + S_3\hat{j}] + C_2[0.5(S_3^2 - S_2^2)\hat{i} + S_2S_3\hat{j}] + C_3[S_1S_2\hat{i} + S_1S_3\hat{j}] +$$

$$C_4[S_2(S_2^2 + S_3^2)\hat{i} + S_3(S_2^2 + S_3^2)\hat{j}] + C_5[0.5S_1(S_3^2 - S_2^2)\hat{i} + S_1S_2S_3\hat{j}] +$$

$$C_6[0.125(S_2^2 - S_3^2 + 2S_2S_3)(S_3^2 - S_2^2 + 2S_2S_3)\hat{i} + S_2S_3(S_3^2 + S_2^2)\hat{j}] +$$

$$C_7[S_1S_2(S_2^2 + S_3^2)\hat{i} + S_1S_3(S_2^2 + S_3^2)\hat{j}] +$$

$$C_8[3.5S_2^5 + 12S_2^3S_3^2 + 2S_2S_3^4)\hat{i} + (3.5S_3^5 + 12S_3^3S_2^2 + 2S_3S_2^4)\hat{j}] +$$

$$C_9[0.125S_1(S_2^2 - S_3^2 + 2S_2S_3)(S_3^2 - S_2^2 + 2S_2S_3)\hat{i} + S_1S_2S_3(S_3^2 + S_2^2)\hat{j}] + ... \quad (30.3)$$

Equation (30.3) includes all cross-terms and so is the highest-order power series expansion reported to date in the fitting of the Li_3^+ dipole moment surface.

Generally in calculating the spectral properties the near rank deficiencies are circumvented by excluding all the S_1 cross-terms, since the inclusion of these expansion variables yield coefficients that are unrealistically large. Large coefficients for high-order terms of a power series expansion is counter to the requirement that the coefficient of successive order terms should be dampened. This is exemplified by comparing entries under columns I and II in Table 8.1. Hence, calculated spectral properties of Li_3^+ [5] and H_3^+ [7] associated with the [6,7] vibrational band origins were rendered as uncertain. However, by applying the SVD analysis (i.e. setting the singular values associated with these expansion variables to zero) retrieves not only the original coefficients, but also yields back-transformed coefficients of the S_1 cross-terms of order 10^{-7} a.u., thereby rationalising their exclusion. Moreover, the difference in χ^2 between the two fits is less than 1%, suggesting that if precision of the fit is an important criterion, the stated apprehensions in the calculated spectral properties involving the [6,7] vibrational band origins of Li_3^+ [5] and H_3^+ [7] can be waylaid (since these properties are strongly dependent on the dipole moment surface). Without SVD analysis such an assertion could not be made with confidence.

Table 8.1 clearly demonstrates that the coefficients of the expansion faithfully follows the linear regions, since the first-order coefficient is numerically the largest coefficient. The fourth and fifth-order coefficients are small and are less accurately determined by the regression analysis. Figure 8.1 shows the constant dipole moment contours in S_2 and S_3 for μ_x and μ_y, with the S_1 value being held at zero. It is evident that both analytical fits of μ_x and μ_y give S_2 and S_3 as zero at the equilibrium geometry and that for both the μ_x and μ_y functions the electrical anharmonicity will cause the intensity to be enhanced (over that predicted by the linear model) for transitions to the highly excited states. However, unlike H_3^+ [7] the calculated saddle points for Li_3^+ [5] occur for only very large amplitudes of vibration. That is, μ_x and μ_y appear linear at the equilibrium geometry ($S_2 = S_3 = 0$) but have saddle points at $S_3 = 0$, $S_2 = 2.12$ au and at $S_3 = 0$, $S_2 = -1.85$ au respectively.

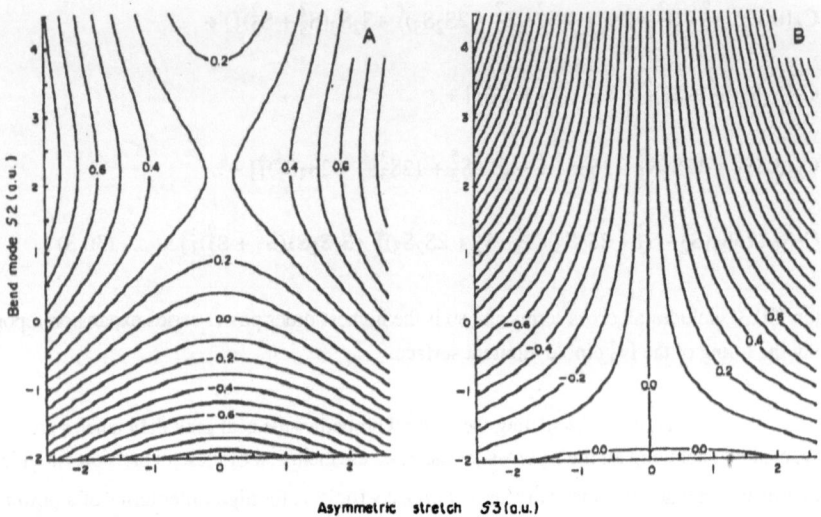

Asymmetric stretch $S3$ (a.u.)

Figure 8.1 A: Constant dipole moment contours for the m_x component (in a.u.) in the S_2-S_3 plane. The symmetric stretch coordinate is fixed at the equilibrium value (i.e. for S_1=0). B: Constant dipole moment contours for the μ_y component (in a.u.) in the S_2-S_3 plane. The symmetric stretch coordinate is fixed at the equilibrium value (i.e. for S_1=0). Reproduced with permission from reference [5].

A variety of ab initio dipole moment surfaces are appearing in the literature [1, 5, 8-9]. For example, a 40 point Hartree-Fock SCF level dipole moment surface was calculated for K_2Li^+ in terms of the rectilinear displacement coordinates [8]. At each data point the geometry was rotated and/or reflected in order to coincide with the Eckart framework. For this charged triatomic molecule of C_{2v} symmetry there is a permanent dipole moment for the μ_x component at the equilibrium geometry, whereas the μ_y component is zero. A dipole moment function was generated from a power series expansion of the rectilinear t coordinates of form [8],

$$\mu_x = C_1 + C_2 (t_1 + t_2) + C_3 (t_1 + t_2)^2 + C_4 t_3^2 + C_5 (t_1 + t_2)^3 + C_6 t_3^2 (t_1 + t_2) +$$
$$C_7 (t_1 + t_2)^4 + C_8 t_3^4 + C_9 t_3^2 (t_1 + t_2)^2 + C_{10} (t_1 + t_2)^5 + C_{11} t_3^4 (t_1 + t_2) +$$
$$C_{12} t_3^2 (t_1 + t_2)^3 + ...$$

$$\qquad (30.4)$$

$$\mu_y = C_1' t_3 + C_2' t_3 (t_1 + t_2) + C_3' t_3^3 + C_4' t_3 (t_1 + t_2)^2 + C_5' t_3^3 (t_1 + t_2) +$$
$$C_6' t_3 (t_1 + t_2)^3 + C_7' t_3^5 + C_8' t_3 (t_1 + t_2)^4 + C_9' t_3^3 (t_1 + t_2)^2 + ...$$

Table 8.2 gives the calculated regression coefficients for K_2Li^+ using the t coordinate expansion. The coefficients clearly show that the non-linear regions are significant since the C_1 coefficients are not necessarily the largest coefficient. However, it should not be forgotten that the fit represents an optimal regression and so reflects its utility as an interpolating function.

§.31. Radiative Properties.

Using rovibrational wavefunctions and a dipole moment surface the intensity of dipole induced transitions can be variationally calculated. The Einstein transition probabilities, A_{nm} (spontaneous emission) and B_{nm} (induced absorption), for transition between energy levels n and m are given by [10],

$$A_{nm} = \frac{32\pi^3 v_{nm}^3}{3\hbar g_n} |\langle \psi_n^{vib} | \mu | \psi_m^{vib} \rangle|^2$$

$$B_{nm} = \frac{2\pi}{3\hbar^2 g_m} |\langle \psi_n^{vib} | \mu | \psi_m^{vib} \rangle|^2 \qquad (31.1)$$

where g_n and g_m are the respective degeneracies.

The vibrational band intensity is given by,

$$S_{nm} = 11.1851 \, v_{nm} \, B_{nm} = 2.3795 \times 10^7 \, f_{nm} \qquad (31.2)$$

where f_{nm} is the oscillator strength, v_{nm} and B_{nm} are in s^{-1} and $erg^{-1} \, s^{-2} cm^3$ respectively with the constant factor converting the left side units to $atm^{-1} \, cm^{-2}$.

The lifetime associated with a vibrational state can be derived by balancing the spontaneous emission with spontaneous and induced absorption [10]. Hence, the lifetime of a vibrational state labelled n is given by,

Table 8.2 Expansion Coefficients for Dipole Moment Surface of K_2Li^{+a}.

Expansion Variable	μ_x	Expansion Variable	μ_y
1	0.69366	t_3	1.10286
$t_1 + t_2$	0.07087	$t_3(t_1 + t_2)$	0.30967
$(t_1 + t_2)^2$	-0.19417	t_3^3	0.06819
t_3^2	0.00227	$t_3(t_1 + t_2)^2$	0.39667
$(t_1 + t_2)^3$	-0.06633	$t_3^3(t_1 + t_2)$	-0.84205
$t_3^2(t_1 + t_2)$	0.48474	$t_3(t_1 + t_2)^3$	-0.23369
$(t_1 + t_2)^4$	0.06298	t_3^5	-0.08589
t_3^4	0.05936	$t_3(t_1 + t_2)^4$	-0.47327
$t_3^2(t_1 + t_2)^2$	-0.29809	$t_3^3(t_1 + t_2)^2$	-2.21253
$(t_1 + t_2)^5$	0.02801		
$t_3^4(t_1 + t_2)$	-0.47670		
$t_3^2(t_1 + t_2)^3$	-1.71098		

a) Reproduced with permission from reference [9]. All entries in atomic units.

$$\tau_n^{-1} = \sum_m A_{nm} \tag{31.3}$$

The square transition dipole matrix elements are given by [11],

$$|R|_{nm}^2 = 3.2737 \times 10^{-4} \, S_{nm} \, Q_v T/(\exp(-E_{v''}/kT) v_0 C) \tag{31.4}$$

where S_{nm} is the vibrational band intensity in $atm^{-1}cm^{-2}$, Q_v is the vibrational partition function, T is the temperature in Kelvin, $E_{v''}$ is the lower state vibrational energy in cm^{-1}, k is Boltzmann constant, v_0 is the vibrational band origin in cm^{-1} and C is the isotopic abundance. The constant factor ensures that the unit of the matrix elements are in D^2.

The absorption line intensities are given by [12],

$$S_i = 3054.6 \, g_{NS} \, v_i \, R^2 \exp(-E_r/kT)(1-\exp(-v_i/kT)/(TQ_r) \tag{31.5}$$

where g_{NS} is the nuclear statistical weight, v_i the transition frequency, E_r is the energy of the rotational level in the vibrational state and Q_r is the rotational partition function. Equation (31.5) already takes into account Hönl-London and Hermann-Wallis factors for symmetric rotors and also the $2J+1$ degeneracy factor since R^2 is calculated exactly [11-12].

Generally Q_r is determined variationally from the rotational levels using [12],

$$Q_r = \sum_m A_{nm} g_{NS} (2J+1) \exp(-E_r/kT) \tag{31.6}$$

However, an independent approximate expression for Q_r is given by [12-13],

$$Q_r = 2[\pi(kT)^3/A_e B_e C_e]^{1/2}$$

where A_e, B_e and C_e are the rotational constants obtained from variationally calculated rotational levels.

For K_2Li^+ the Einstein transition probabilities, band strengths and vibrational radiative lifetimes have been calculated using vibrational eigenfunctions and eigenenergies [8] and the dipole moment function detailed in Table 8.2. These quantities are given in Table 8.3 with respect to transitions from the ground vibrational state to the ninth excited vibrational state. For molecules

Table 8.3 Vibrational Transition Frequencies, Square Dipole Matrix Elements, Einstein Coefficients, Band Strengths and Radiative Lifetimes for K_2Li^{+a}.

j	i	ν_{ji} /cm^{-1}	μ_{ji}^2 /D^2	A_{ji} /s^{-1}	B_{ji} /10^{16}cm^{3}erg^{-1}s^{-1}	S_{ji} /atm^{-1}cm^{-2}	τ^b /s
1	0	23.90	0.058	0.0000	0.6433	0.9135	0.14
2	0	55.63	0.002	0.0000	0.0007	0.0023	0.10
3	0	93.74	0.150	0.0056	4.0816	22.7301	0.22
4	0	140.82	0.260	0.0595	12.7841	106.9510	0.38
5	0	166.04	0.370	0.1946	25.5239	251.7725	1.25
6	0	166.18	0.017	0.0004	0.0545	0.5380	11.82
7	0	189.97	0.100	0.0220	1.9248	21.7230	1.13
8	0	190.42	0.240	0.1264	10.9940	124.3672	1.20
9	0	193.27	0.110	0.0259	2.1553	24.7470	3.33

a) Reproduced with permission from reference [9]. Square dipole matrix elements and band strengths calculated at 300 K.

b) Lifetime of the upper state.

belonging to the C_{2v} point group there are no Raman forbidden transitions, unlike Li_3^+ which has D_{3h} symmetry. Hence the lifetimes of the excited vibrational states of K_2Li^+ are small for these transitions compared with those of Li_3^+ [5]. Nevertheless, the longest lifetime 11.82 seconds is for an excited state with an assigned principal configurations of (010) and (001). Similarly, for Li_3^+ the first excited state with principal configuration assigned as (010) has calculated lifetime of 12.34 seconds [5].

The rotational partition function was obtained from the variationally calculated rotational levels in order to determine the rotational line intensities. The intensities were calculated at 300 K, with contributions from the vibrational partition function being neglected. Table 8.4 gives the absolute line intensities and the squares of the electric dipole transition matrix elements for some of the most intense transitions within the P,Q and R branches between the vibrational ground state and lowest-lying four excited vibrational states. At present there is no rovibrational experimental data available on this molecule and so these calculations are speculative.

REFERENCES TO CHAPTER VIII

1 Dykstra CE (1988) Ab initio calculation of the structures and properties of molecules, Elsevier, New York
2 Papoušek D, Aliev MR (1982) Molecular vibrational-rotational spectra, Elsevier, New York
3 Wang F, Searles DJ, von Nagy-Felsobuki EI (1992) J Chin Chem Soc 39:339
4 Green S (1974) Adv Chem Phys 25:179
5 Searles DJ, Dunne SJ, von Nagy-Felsobuki EI (1988) Spectrochim Acta A44: 985
6 Gerber WH and Schumacher E (1978) J Chem Phys 69:1692
7 Carney GD, Porter RN (1976) J Chem Phys 65:3547
8 Searles DJ, von Nagy-Felsobuki, EI (1991) In: Vibrational spectra and structure, Durig JR (Ed) , Vol 19, Elsevier, Amsterdam
9 Wang F, Searles DJ, von Nagy-Felsobuki EI (1992) J Phys Chem 96:6158
10 Penner SS (1959) Quantitative molecular spectroscopy and gas emissivities, Addison-Wesley, Reading
11 Johns JWC (1987) J Mol Spectrosc 125:442
12 Weis B, Carter S, Rosmus P, Werner H-J, Knowles PJ (1989) J Chem Phys 91:2818
13 Strow LL (1983) J Quant Spectrosc Radiat Transf 29:395

Table 8.4 Variationally Calculated Rovibrational Absorption Line Intensities (at 300 K) for χ States of K_2Li^{+a}.

v'	J'	K_a'K_c'		v"	J"	K_a"K_c"		Branch P,Q,R	v_{AX} [b]	S_{AX} [c]	R_{AX}^2 [d]
0	5	3	2	0	6	0	6	-1	3.18	0.370-4	0.180+1
0	3	3	0	0	3	1	2	0	3.12	0.475-4	0.238+1
0	5	4	2	0	4	3	1	1	3.02	0.409-4	0.223+1
1	5	2	4	0	6	2	4	-1	23.56	0.329-5	0.307-2
1	7	7	0	0	7	1	6	0	42.49	0.113-4	0.337-2
1	6	5	1	0	5	2	4	1	32.39	0.223-6	0.112-3
3	3	2	2	0	4	3	1	-1	91.53	0.263-3	0.194-1
3	4	0	4	0	4	1	3	0	93.36	0.273-3	0.189-1
3	5	3	2	0	4	2	2	1	95.93	0.303-3	0.201-1
4	4	3	2	0	5	1	4	-1	143.56	0.207-2	0.675-1
4	5	1	4	0	5	2	3	0	139.69	0.195-2	0.670-1
4	5	4	2	0	4	1	4	1	146.83	0.317-3	0.996-2

a) Reproduced with permission from reference [9].

b) Difference of eigenenergies between rovibrational states labelled as A and X, where A and X represent the upper and lower rovibrational states respectively. All entries in units of cm^{-1}.

c) Band strengths between rovibrational states labelled as A and X, where A and X represent the upper and lower rovibrational states respectively. All entries in units of $atm^{-1}cm^{-2}$.

d) Square of the dipole matrix element spanned by the rovibrational states labelled as A and X, where A and X represent the upper and lower rovibrational states respectively. All entries in units of D^2.

CHAPTER IX

APPLICATIONS TO BENT TRIATOMIC MOLECULES

§.32. Introduction.

For electron-sparse molecules, ab initio electronic PE surfaces of spectroscopic quality are readily available within the framework of the Born-Oppenheimer approximation [1-2]. For more electron-dense molecules (such as Li_3^+) speculative configuration interaction PE surfaces have been constructed because of restrictions on computing capacity [3-4]. While the latter surfaces are primitive, they nevertheless serve experimentalists and theoreticians alike in the exploration of what constitutes a cost-effective, but spectroscopically predictive ab initio PE surface.

Ab initio discrete PE surfaces and subsequent rovibrational calculations of the alkali metal cations are of prime importance due to their predicted role in a number of technologically important processes (such as lamps for high-resolution optically pumped lasers) [5]. Surprisingly, these singly charged triatomic cations are more strongly bound than their neutral analogues. For example, the Li_3 CEPA binding energy [6] is 1.47 eV, whereas for Li_3^+ a pseudopotential-CI calculation [7] predicts the binding energy to be 1.77 eV. For Li_2Na^+ the pseudopotential-CI calculation gives the binding energy to be 1.1 eV larger than the corresponding neutral molecule [7].

As has been pointed out by Herzberg [8], perhaps one of the most important triatomic ions is H_2O^+, which is found in a variety of environments such as in the earth's atmosphere, interstellar space and oxygen-rich circumstellar envelopes. Lew [9], Strahan et al. [10] and Dinelli et al. [11] have reported high resolution spectra for the vibrational ground state, the v_1, $2v_2$ and $3v_2$ and v_3 vibrational states for the electronic ground state ($\chi\ ^2B_1$) of this ion. For D_2O^+, Lew and Groleau [12] have reported data on the vibrational ground state, the v_2 and $3v_3$ vibrational states. Karlsson et al. [13], using photoelectron spectroscopy have provided data on the higher overtones of H_2O^+, D_2O^+ and HDO^+.

The rovibrational calculations of Na_3^+ [14], H_2O^+ [15] and $KLiNa^+$ [16-17] are of interest to the spectroscopic community. These molecules exhibit use of the **s**, **t** and **f** coordinate rovibrational Hamiltonians (see Chapter VI) and are a severe test of the solution algorithm. Moreover, in the case of Na_3^+ and H_2O^+ comparison can be made with other variational calculations using Hamiltonians cast in different coordinate systems and furthermore, in the case of H_2O^+ comparisons can be made with experiment.

§.33. D$_{3h}$ Case: Na$_3^+$

The electrical conductivity of sodium vapour above 10 torr must take into account the presence of ions such as Na$_3^+$ [5]. Experimentally, Na$_3^+$ has been identified in electron and photoionization mass spectra of supersonic molecular beams of sodium metal vapours [18-19]. It is chemically stable and so cannot be considered as being held together by weak van der Waals forces. Furthermore, its geometry is of D$_{3h}$ symmetry with the charge being at the centroid of an equilateral triangle [20].

Ab initio electronic calculations were performed using the GAUSSIAN 88 suite of programmes [21] within the SDCI ansatz and using the frozen-core (FC) approximation. The basis set used for sodium was Huzinaga et al. [22] [16s9p] primitive basis which was supplemented with diffuse d and f polarization functions (both with a partially optimized exponent of 0.16) yielding a [16s9p1d1f/10s6p1d1f] contracted basis.

The SDCI/FC model predicts Na$_3^+$ to be of D$_{3h}$ symmetry with an equilibrium bond length of 3.511 Å. This is in excellent agreement with the pseudopotential-CI calculation of Carter and Meyer [23], which yields an equilibrium bond length of 3.576 Å. The SDCI/FC minimum energy is -485.4679 E$_h$, whereas the pseudopotential-CI minimum energy is -485.4409 E$_h$. This disparity may not reflect the extent of valence correlation recovery, since for a many valence electron system the core-core and core-valence electronic interactions cannot be rigourously modelled in the pseudopotential formalism and so that the total energy depends on the definition of valence and core sub-spaces. Hence comparisons of total energies are inappropriate.

A 61 point discrete SDCI/FC PE surface constructed by using an iterative procedure [14], which ensures that the final energy grid yields calculated electronic energies close to the quadrature points chosen by the potential energy integrator. Furthermore, additional points were selected within this grid to "better" define the analytical representation in regions of "poor" fit.

Even though the grid is weighted toward the quadrature points it is nevertheless important to obtain force fields that accurately interpolate the surface between calculated points. For Na$_3^+$ the "best" fit to the discrete surface is a Padé approximates expansion, denoted as P(4,6), with a Dunham expansion variable. It is necessary to set singular values σ_{29-30} to zero in order to ensure that the analytical PE surface is free from singularities in the integration region. Table 9.1 gives the analytical representation of the PE surface which was used in the rovibrational calculations.

Table 9.1 Expansions Coefficients for the P(4,6) PE Surface of Na_3^{+a}.

Dunham Expansion Parameter[b]	Numerator	Denominator
1	-485.467905	1.000000
$\rho_1 + \rho_2 + \rho_3$	-371.534931	-0.765313
$\rho_1^2+\rho_2^2+\rho_3^2$	-432.582752	-0.891336
$\rho_1\rho_2 + \rho_2\rho_3 + \rho_1\rho_3$	-1895.095330	-3.903647
$\rho_1^3 + \rho_2^3 + \rho_3^3$	166.597480	0.343552
$\rho_1^2(\rho_2+\rho_3) + \rho_2^2(\rho_1+\rho_3) + \rho_3^2(\rho_1+\rho_2)$	-4415.484260	-9.095503
$\rho_1\rho_2\rho_3$	-9425.828479	-19.416159
$\rho_1^4 + \rho_2^4 + \rho_3^4$	178.768603	0.367752
$\rho_1^3(\rho_2+\rho_3) + \rho_2^3(\rho_1+\rho_3) + \rho_3^3(\rho_1+\rho_2)$	-1894.607809	-3.903213
$\rho_1^2\rho_2^2 + \rho_1^2\rho_3^2 + \rho_2^2\rho_3^2$	-3224.710355	-6.643138
$\rho_1\rho_2\rho_3^2 + \rho_3\rho_2\rho_1^2 + \rho_3\rho_1\rho_2^2$	-5859.658959	-12.071164
$\rho_1^5+\rho_2^5+\rho_3^5$	-	0.000538
$\rho_1(\rho_2^4 + \rho_3^4)+\rho_2(\rho_1^4 + \rho_3^4) +\rho_3(\rho_1^4 + \rho_2^4)$	-	-0.000619
$\rho_1^3(\rho_2^2 + \rho_3^2)+\rho_2^3(\rho_1^2 + \rho_3^2) +\rho_3^3(\rho_1^2 + \rho_2^2)$	-	-0.001861
$\rho_1\rho_2\rho_3^3 + \rho_2\rho_3\rho_1^3 + \rho_3\rho_1\rho_2^3$	-	-0.002664
$\rho_1\rho_2^2\rho_3^2 + \rho_2\rho_3\rho_1^2 + \rho_3\rho_1^2\rho_2^2$	-	-0.005679
$\rho_1^6 + \rho_2^6 + \rho_3^6$	-	-0.000265
$\rho_1^5(\rho_2+\rho_3) + \rho_2^5(\rho_3+\rho_1) + \rho_3^5(\rho_2+\rho_3)$	-	0.000811
$\rho_1^4(\rho_2^2 + \rho_3^2) + \rho_2^4(\rho_1^2 + \rho_3^2) + \rho_3^4(\rho_1^2 + \rho_2^2)$	-	0.002577
$\rho_1^4\rho_2\rho_3 + \rho_2^4\rho_1\rho_3 + \rho_1\rho_2\rho_3^4$	-	0.003627
$\rho_1^3\rho_2^3 + \rho_1^3\rho_3^3 + \rho_2^3\rho_3^3$	-	-0.002647
$\rho_1^3\rho_2\rho_3(\rho_2 + \rho_3) + \rho_1\rho_2^3\rho_3(\rho_1 + \rho_3)$ $+ \rho_1\rho_2\rho_3^3(\rho_1 + \rho_2)$	-	0.000730
$\rho_1^2\rho_2^2\rho_3^2$	-	-0.003067
$(\chi^2)^{1/2}$	3.45 x 10^{-5}	

a) Reproduced with permission from reference [14].

b) See Chapter IV for an explanation of the expansion variable.

The vibrational Hamiltonian used was the s coordinate Hamiltonian, utilising a third-order Taylor series expansion of the Watson operator (see Chapter VI). It has the full mechanical anharmonicity embedded as well as the operators which couple the vibrational modes. For each coordinate, 1000 finite-elements were constructed within the following domains: s_1 [-2.2 a.u., 2.8 a.u.], s_2 [-2.8 a.u., 3.0 a.u.], s_3 [-2.8 a.u., 2.8 a.u.]. A three-dimensional configuration basis was spliced from the one-dimensional solutions with the configuration list selected using a nodal cut-off criterion (see Chapter VII). The secular determinant was constructed and diagonalised to yield variational vibration wavefunctions and eigenenergies, both of which are required for the rovibrational problem.

Table 9.2 assigns the 20 lowest-lying vibrational band origins of Na_3^+. The vibrational states have been labelled using the v_2 quantum number, with members of the v_2 group being assigned using a second quantum number l_2, where $l_2 = -v_2, -v_2+2, ..., v_2$. For the D_{3h} symmetry, levels with $l_2 = 0$ are assigned to A_1', those with $l_2 = \pm 3, \pm 6$ etc. to A_1', A_2' pairs and all others to E. The vibrational states have been assigned using the weighted coefficients of the basis functions. The D_{3h} symmetry of the molecule yields a simple assignment, since mixing can only occur for configurational basis functions belonging to the same irreducible representation. Table 9.2 emphasises this point by giving the percentage weight of the dominant configuration basis functions and highlights the vibrational band origins up to 204 cm^{-1}.

The agreement between the assignment given by Table 9.2 and those of Carter and Meyer [23] are excellent, with no apparent cross-overs within the sequence. The maximum difference between the two sets of calculations is 9 cm^{-1} and that is for the higher-lying vibrational band origins. For the lower-lying vibrational band origins differences are between 3-4 cm^{-1}. Moreover, Table 9.3 compares the zero-order vibrational frequencies and anharmonic constants obtained using the vibrational eigenenergies from both calculations. These constants further demonstrate that the level of agreement between the two calculations are excellent. It should not be forgotten that this level of agreement has been reached by using two vastly different variational solution algorithms. The Carter and Meyer [23] solution algorithm relies on a discrete pseudopotential-CI surface with an analytical potential energy function expanded in terms of Morse displacement coordinates, whereas Wang and von Nagy-Felsobuki [14] have employed an ab initio all-electron SDCI discrete surface using a P(4,6) force field. Furthermore, the Carter and Meyer [23] vibrational Hamiltonian is based on a hyperspherical coordinates system, whereas Wang and von Nagy-Felsobuki [14] have utilised the s coordinate Hamiltonian. Although Carter and Meyer [23] have employed the same potential integrator their solution algorithm is reliant upon analytical basis functions, whereas Wang and von Nagy-Felsobuki [14] solution algorithm is solely dependent on numerical basis functions (see Chapter V).

Table 9.2 Assigned Vibrational Band Origins of Na_3^+ ($/cm^{-1}$)[a].

$v_A v_E l/2$	Symmetry	% weight	Vibrational Band Origin	
			Wang and Felsobuki[b]	Carter and Meyer[c]
011	E'	99	101.9	100.0
		99	101.9	100.0
100	A_1'	98	143.8	140.5
020	A_1'	48, 47	202.8	198.9
022	E'	91	203.6	199.7
		47, 46	203.6	199.7
111	E'	95	244.7	239.3
		95	244.7	239.3
200	A_1'	97	287.1	280.4
031	E'	70, 22	303.6	297.7
		67, 22	303.6	297.7
033	A_1'	68, 25	305.3	299.3
033	A_2'	71, 24	305.4	299.4
120	A_1'	44, 43	344.6	337.2
122	E'	46, 45	345.4	337.9
		90	345.4	337.9
211	E'	92	386.9	378.1
		91	386.9	378.1
040	A_1'	33 , 31	403.5	395.4

a) Reproduced with permission from reference [14].

b) The zero-point energy of Na_3^+ is 174.61 cm^{-1}. See reference [14].

c) See reference [23].

Table 9.3 Zero-Order Vibrational Frequencies and Anharmonic Constants of Na_3^+ ($/cm^{-1}$)[a].

Constant	Wang and Felsobuki[b]	Carter and Meyer[c]
ω_1	145.3233	142.09101
ω_2	102.9447	101.18739
χ_{11}	-0.2599	-0.29961
χ_{22}	-0.2614	-0.29875
χ_{12}	-1.0047	-1.06633
γ_{22}	0.2089	0.19677

a) Reproduced with permission from reference [14].
b) See reference [14].
c) See reference [23].

For a D_{3h} molecule, such as for the most abundant isotope of Na_3^+, with three nuclei of spin I=3/2, nuclear spin functions with A_1', A_2' and E' symmetry are possible. Hence the totally symmetric rotationless zero-point vibrational level (i.e. J=0, K=0 state) and the two lowest excited rotational states (i.e. J=1, K=1 state of E" symmetry and the J=1, K=0 state of A_2' symmetry) are accessible, since each of these states can couple with nuclear spin states and so ensure that the total wavefunction is antisymmetric, thereby not violating the Pauli's exclusion principle for Fermions. Hence all rotational levels can occur. Table 9.4 gives the variationally calculated rovibrational eigenenergies (up to J=5) for the seven lowest-lying vibrational states. In order to ensure convergence of the calculated eigenenergies with respect to the basis set used, a number of truncated basis sets were investigated. Calculations were employed using seven and fifteen vibrational eigenfunctions respectively. The mean difference of all the rotational levels of the first five vibrational states using these two different vibrational basis sets is 0.001 cm^{-1}.

High resolution rovibrational spectra are usually assigned by fitting rovibrational data to reduced Hamiltonians (see Chapter). Using the ab initio rovibrational states, the fitted rotational constants A_0, A_e, C_0 and C_e are 0.1213, 0.1216, 0.0604, 0.0607 cm^{-1} respectively [14]. The fitted coefficients α_1^A, α_2^A, α_1^C and α_2^C are 0.0003, -0.0001, 0.0002 and 0.0002 cm^{-1} respectively [14].

Table 9.4 Rotational Energy Levels for Low-Lying Vibrational States of Na_3^+ ($/cm^{-1}$)[a].

E_v	0.000	101.862[b]	143.796	202.796	203.639[b]
J K					
1 1	0.182	0.182	0.181	0.181	0.181
1 0	0.243	0.243	0.242	0.243	0.243
2 2	0.484	0.483	0.483	0.483	0.483
2 1	0.667	0.667	0.665	0.667	0.666
2 0	0.728	0.728	0.726	0.728	0.728
3 3	0.908	0.906	0.905	0.904	0.904
3 2	1.212	1.211	1.208	1.210	1.210
3 1	1.394	1.394	1.390	1.394	1.394
3 0	1.455	1.455	1.451	1.456	1.556
4 4	1.452	1.449	1.448	1.445	1.508
4 3	1.878	1.876	1.872	1.874	1.874
4 2	2.182	2.181	2.176	2.181	2.180
4 1	2.364	2.364	2.358	2.302	2.365
4 0	2.425	2.426	2.419	2.302	2.426
5 5	2.117	2.112	2.111	2.107	2.628
5 4	2.664	2.661	2.657	2.658	2.721
5 3	3.090	2.089	3.082	2.963	3.087
5 2	3.394	3.394	3.385	2.963	3.393
5 1	3.577	3.577	3.567	3.087	3.578
5 0	3.638	3.639	3.628	3.394	3.638

a) Reproduced with permission from reference [14].

b) Only one degenerate component listed.

§.34. C$_{2v}$ Case: H$_2$O$^+$

Weis et al. [2] have theoretically resolved the infrared spectrum of H$_2$O$^+$, D$_2$O$^+$ and HDO$^+$ using the "complete" approach. That is, they have calculated a multiconfiguration reference configuration interaction (MRCI) discrete potential energy surface and embedded an analytical representation of it in an internal coordinate rovibrational Hamiltonian, which they have variationally solved using basis functions constructed from Morse, harmonic oscillator and associate Legendre functions. While the χ ^{2}B$_1$ and A ^{2}A$_1$ of H$_2$O$^+$ ion form a Renner-Teller pair, it is nevertheless possible to treat the electronic ground state potential energy function as a single state problem near the potential energy minimum. Generally Renner-Teller coupling effects need only to be included in regions of the potential close to linear structures.

Although Wang and von Nagy-Felsobuki [15] have employed the discrete MRCI potential of Wies et al. [2] they have utilised a different force field embedded within the Eckart-Watson Hamiltonian and have used different rovibrational trial functions in their variational solution. Nevertheless, t coordinate solution produces rovibrational eigenenergies in excellent agreement with the internal coordinate solution of Weis et al. [2] and with experiment for the isotopes (where comparison can be made). Furthermore, they have determined [15] rovibrational states for the C$_{2v}$ isotopes where no experimental or ab initio data are currently available.

Weis et al. [2] have calculated an extensive discrete PE surface using highly correlated MRCI electronic wavefunctions. Their analytical representation of the 50 point surface involved fitting a polynomial expansion in Simon-Parr-Finlan bond stretching coordinates and a cubic expansion in terms of an angle bending coordinate. Wang and von Nagy-Felsobuki [15] have refitted this discrete surface using Padé approximates. The coefficients of the fit are given in Table 9.5.and were obtained using the SVD analysis [15].

Table 9.5 shows that the root-mean-square error of the fit is 2.6×10^{-5} E$_h$. Numerous graphical inspections of the surface were performed in order to ensure that no oscillatory behaviour or singularities were presented in the potential energy integration region. For example, Wang and von Nagy-Felsobuki [15] did obtain a P(3, 6) surface with a root-mean-square error of 6.3×10^{-6} E$_h$. However, on graphical inspection it was found to have singularities present in the t_2-t_3 integration region. Hence such a surface is not suitable for embedding in the nuclear Schrödinger equation. However, by zeroing an additional singular value σ_{45}, an artifact free surface was obtained, even though the root-mean-square error of the fit is one magnitude larger. Figure 9.1 gives a contour plot of the potential energy surface, showing that the surface contains no singularities.

Table 9.5 Expansion Coefficients for the P(3, 6) Power Series Representation of the PE Surfaces of H_2O^+ [a].

		H_2O^+	
	Expansion Variable	Numerator	Denominator
1	1	-75.88840	1.00000
2	$\rho_1 + \rho_2$	-0.00289	0.00009
3	ρ_3	0.00758	0.00003
4	$\rho_1^2 + \rho_2^2$	0.00167	-0.01123
5	ρ_3^2	0.00281	-0.00745
6	$\rho_1\rho_2$	0.00148	-0.00254
7	$\rho_2\rho_3 + \rho_1\rho_3$	--0.00070	0.00619
8	$\rho_1^3 + \rho_2^3$	-0.00032	0.01424
9	ρ_3^3	0.00149	-0.01355
10	$\rho_1^2\rho_2 + \rho_2^2\rho_1$	-0.00090	0.00695
11	$\rho_1^2\rho_3 + \rho_2^2\rho_3$	0.00071	-0.01012
12	$\rho_1\rho_3^2 + \rho_2\rho_3^2$	-0.00014	0.02324
13	$\rho_1\rho_2\rho_3$	0.00085	-0.02884
14	$\rho_1^4 + \rho_2^4$	0.00000	-0.00909
15	ρ_3^4	0.00000	-0.02489
16	$\rho_1^3\rho_2 + \rho_2^3\rho_1$	0.00000	-0.00498
17	$\rho_1^3\rho_3 + \rho_2^3\rho_3$	0.00000	0.00587
18	$\rho_1\rho_3^3 + \rho_2\rho_3^3$	0.00000	0.04830
19	$\rho_1^2\rho_2^2$	0.00000	-0.01122
20	$\rho_1^2\rho_3^2 + \rho_2^2\rho_3^2$	0.00000	-0.02896
21	$\rho_1^2\rho_2\rho_3 + \rho_1\rho_2^2\rho_3$	0.00000	0.03843
22	$\rho_1\rho_2\rho_3^2$	0.00000	-0.07975
23	$\rho_1^5 + \rho_2^5$	0.00000	0.00118
24	ρ_3^5	0.00000	-0.06241
25	$\rho_1^4\rho_2 + \rho_2^4\rho_1$	0.00000	-0.01075
26	$\rho_1^4\rho_3 + \rho_2^4\rho_3$	0.00000	0.01741
27	$\rho_1\rho_3^4 + \rho_2\rho_3^4$	0.00000	0.10051
28	$\rho_1^3\rho_2^2 + \rho_1^2\rho_2^3$	0.00000	-0.01335
29	$\rho_1^3\rho_3^2 + \rho_2^3\rho_3^2$	0.00000	-0.02313
30	$\rho_1^2\rho_3^3 + \rho_2^2\rho_3^3$	0.00000	-0.03330
31	$\rho_1^3\rho_2\rho_3 + \rho_1\rho_2^3\rho_3$	0.00000	0.03192

Table 9.5 (Cont.)

32	$\rho_1\rho_2\rho_3^3$	0.00000	-0.15018
33	$\rho_1^2\rho_2^2\rho_3$	0.00000	0.02214
34	$\rho_1^2\rho_2\rho_3^2 + \rho_1\rho_2^2\rho_3^2$	0.00000	0.02854
35	$\rho_1^6 + \rho_2^6$	0.00000	0.00040
36	ρ_3^6	0.00000	-0.11660
37	$\rho_1^5\rho_2 + \rho_2^5\rho_1$	0.00000	-0.00020
38	$\rho_1^5\rho_3 + \rho_2^5\rho_3$	0.00000	-0.01429
39	$\rho_1\rho_3^5 + \rho_2\rho_3^5$	0.00000	0.19406
40	$\rho_1^4\rho_2^2 + \rho_2^4\rho_1^2$	0.00000	-0.05428
41	$\rho_1^4\rho_3^2 + \rho_2^4\rho_3^2$	0.00000	0.03911
42	$\rho_1^2\rho_3^4 + \rho_2^2\rho_3^4$	0.00000	-0.09080
43	$\rho_1^4\rho_2\rho_3 + \rho_2^4\rho_1\rho_3$	0.00000	0.04773
44	$\rho_1\rho_2\rho_3^4$	0.00000	-0.44941
45	$\rho_1^3\rho_2^3$	0.00000	-0.08927
46	$\rho_1^3\rho_3^3 + \rho_2^3\rho_3^3$	0.00000	-0.03002
47	$\rho_1^3\rho_2^2\rho_3 + \rho_1^2\rho_2^3\rho_3$	0.00000	0.33846
48	$\rho_1^3\rho_2\rho_3^2 + \rho_1\rho_2^3\rho_3^2$	0.00000	-0.20638
49	$\rho_1^2\rho_2\rho_3^3 + \rho_1\rho_2^2\rho_3^3$	0.00000	0.41913
50	$\rho_1^2\rho_2^2\rho_3^2$	0.00000	-0.64414

$(\chi^2)^{1/2}/E_h$	2.6×10^{-5}

a) Reproduced with permission from reference [15]. Units of all coefficients are in a.u. A full description of Padé approximates and notation used is given in Chapter IV. The singular values σ_{45}, σ_{49}, σ_{51}-σ_{60} were zeroed in order to ensure no oscillatory behaviour or singularities in the integration region.

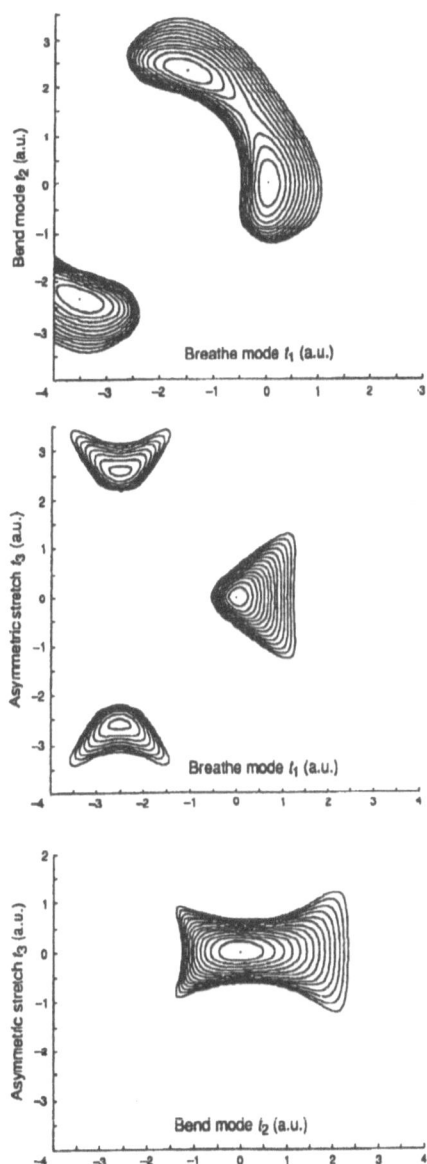

Figure 9.1 Two-dimensional constant potential energy plots for the P(3,6) surface of H_2O^+ with SVD analysis. Reproduced with permission from reference[15].

The P(3,6) analytical representation accurately mimics experimental structures near equilibrium geometry. The experimental equilibrium O-H bond length and H-O-H bond angle, given by Dinelli et al.[11], are 1.001 Å and 108.9° respectively, which are in excellent agreement with P(3, 6) fitted result of 1.0004 Å and 109.07° respectively. Both are in excellent agreement with fitted result of Weis et al.[2], who obtained values of 1.0004 Å and 109.06° respectively. Furthermore, both the theoretically calculated r_0 and α_0 structures agree to within 0.002 Å and 0.2° with the experimental structures of Lew [9] and Dinelli et al.[11].

The theoretical rovibrational states for the ground electronic state $\chi\,^2B_1$ of H_2O^+ have been calculated using t coordinate rovibrational Hamiltonian takes into full account mechanical anharmonicity as well as vibration-rotation coupling effects. However, nuclear spin interactions, the electron spin-rotation and the electronic angular momentum-rotation (commonly referred to as the Renner-Teller effect) have been neglected in the calculations. The one-dimensional wavefunctions are calculated using a finite-element solution of a one dimensional Hamiltonian, which is expressed in terms of a single rectilinear t coordinate. In the one-dimensional Hamiltonian only the second-order expansion of the Watson operator was used, since its is diagonal in the t coordinates. For each t coordinate, 1000 finite-elements were constructed within the following domains: t_1 [-1.7, 3.5], t_2 [-2.5, 4.5], t_3 [-2.0, 2.0]. A three-dimensional configuration basis was spliced from 20x20x20 one-dimensional eigenfunctions. The configuration list was selected in the usual manner.

Table 9.6 gives the zero-point energies and assignments of the lowest-lying vibrational band origins. Care must be taken with zero-point energies, since the centre of global expansion may be below the PE minimum. Nevertheless, from the errors associated with the fit the zero-point energies are expected to be accurate to within ±6 cm^{-1}.

Table 9.7 compares the lowest experimental vibrational band origins of H_2O^+ and D_2O^+ (where available) with ab initio calculated variational and perturbation theories. The perturbation theory employed by Weis et al.[2] includes the Darling-Dennison corrections. All calculations are in essentially good agreement with the available experimental data even though the variational approach employed in this work is vastly different in form and substance from that used by Weis et al. [2]. For H_2O^+ the maximum difference between the experimental data and the t coordinate and internal coordinate solutions is 44 and 135 cm^{-1} respectively. Similarly for D_2O^+ the maximum error is 25 and 19 cm^{-1} respectively. Although agreement between the t coordinate and internal coordinate variational solutions is in particular good for D_2O^+, the major discrepancy between these as well as the perturbation calculations for both H_2O^+ and D_2O^+ arises from states highly excited in the bending mode. The bend potential is shallow and so states that are highly excited in the bending mode are extremely sensitive to basis set incompleteness. With respect to our solution algorithm one would need to incorporate a total list far greater than 560 basis functions and

Table 9.6. Zero-Point Energies (ZPE), Vibrational Band Origins (in cm^{-1}) and Assignment of the C$_{2v}$ Isotopes of H$_2$O^{+a}.

$(v_1v_2v_3)^b$	H$_2^{16}$O$^+$	D$_2^{16}$O$^+$	T$_2^{16}$O$^+$	H$_2^{18}$O$^+$	D$_2^{18}$O$^+$	T$_2^{18}$O$^+$
ZPE	4096.5	3004.1	2536.5	4083.1	2995.0	2514.8
(010)	1412.1	1047.5	888.7	1406.3	1043.2	880.5
(020)	2780.7	2071.9	1760.9	2769.4	2063.4	1744.9
(100)	3212.5	2346.6	1968.6	3206.4	2345.0	1957.3
(001)	3255.0	2424.4	2067.2	3241.8	2412.9	2045.0
(030)	4098.5	3070.9	2615.3	4082.4	3058.5	2591.8
(110)	4611.4	3385.8	2851.2	4599.5	3379.8	2831.7
(011)	4641.5	3458.0	2945.9	4622.6	3442.2	2915.7
(040)	5353.9	4042.0	3450.8	5333.8	4026.0	3420.2
(120)	5970.8	4403.7	3718.4	5953.6	4393.6	3691.0
(021)	5988.1	4469.8	3808.9	5963.8	4449.9	3771.0
(200)	6300.0	4641.9	3904.4	6284.6	4637.4	3881.8
(101)	6312.2	4689.2	3978.9	6293.6	4676.1	3946.0
(002)	6476.4	4812.2	4103.1	6454.6	4791.4	4060.3
(050)	6529.5	4981.2	4265.7	6507.5	4962.1	4228.5
(130)	7278.1	5397.0	4568.1	7256.8	5383.1	4533.5
(031)	7288.0	5457.7	4654.9	7258.7	5433.9	4609.7
(060)	7631.2	5674.7	4782.9	7612.4	5665.2	4751.8
(210)	7681.3	5715.0	4851.9	7658.8	5697.4	4810.8
(111)	7686.8	5833.6	4972.1	7662.3	5808.7	4921.6

a) Reproduced with permission from reference [15].

b) Vibrational assignment is based on the weights of the coefficients (see Chapter VII).

Table 9.7 Comparison of Experimental with Ab Initio Vibrational Band Origins (in cm^{-1}) of $H_2{}^{16}O^+$ and $D_2{}^{16}O^+$ [a].

	Exp.[b]	Wang and Felsobuki[c]	Weis et al.[2]	
			Var.[d]	Pert.[d]
		(a) $H_2{}^{16}O^+$		
(010)	1408.4	1412.1	1410.4	1413.1
(020)	2771.3	2780.7	2774.2	2786.2
(100)	3213.0	3212.5	3211.0	3208.1
(001)	3259.0	3255.0	3255.0	3251.3
(030)	4085.0	4098.5	4083.6	4119.2
(110)	4593.0	4611.4	4600.1	4600.9
(011)	-	4641.5	4639.9	4637.9
(040)	-	5353.9	5323.4	5412.1
(120)	5936.0	5970.8	5942.3	5953.6
(021)	-	5988.1	5980.8	5984.5
(200)	6298.0	6300.0	6296.7	6282.2
(101)	-	6312.2	6310.2	6295.8
(002)	-	6476.4	6475.0	6466.2
(050)	-	6529.5	6462.9	6665.0
(130)	7234.0	7278.1	7228.6	7266.2
(031)	-	7288.0	7271.2	7291.0
(060)	7639.0	7631.2	7496.6.2	7650.9
(210)	-	7681.3	7662.2	7662.0
(111)	-	7686.8	7672.9	7830.1

Table 9.7 (Cont.)

(b) $D_2{}^{16}O^+$

(010)	1044.3	1047.5	1047.0	1048.1
(020)	2063.0	2071.9	2070.2	2074.5
(100)	2342.0	2346.6	2346.2	2345.2
(001)	-	2424.4	2424.2	2422.7
(030)	3058.7	3070.9	3067.5	3079.2
(110)	3373.0	3385.8	3382.2	3383.0
(011)	-	3458.0	3457.8	3457.0
(040)	-	4042.0	4035.8	4062.2
(120)	4396.0	4403.7	4393.6	4399.1
(021)	-	4469.8	4468.4	4469.6
(200)	4638.0	4641.9	4640.7	4636.0
(101)	-	4689.2	4688.5	4683.1
(002)	-	4812.2	4812.5	4807.5
(050)	-	4981.2	4970.5	5023.5
(130)	-	5397.0	5378.5	5393.5
(031)	-	5457.7	5454.4	5460.5
(060){210}[e]	5650.0	5674.7	5666.1	5662.9
(210){111}[e]	-	5715.0	5710.8	5707.1
(111){012}[e]	-	5833.6	5833.3	5828.5

a) Reproduced with permission from reference [15].

b) See Lew [9], Lew and Groleau [12] and Strahan et al. [10].

c) Variational solution. See reference [15].

d) Variational and perturbation solution as given by Weis et al. [2].

e) Weis et al.[2] assignment is given in braces.

in particular a configuration list incorporating more t_2 functions. However, as will be shown below the **t** coordinate approach reproduces rovibrational transition energies to high accuracy when compared with experiment for the ground and first excited vibrational states. For these vibrational states both variational calculations are in excellent agreement and so have properly converged with respect to the basis set expansions.

Both variational calculations give the same assignment for H_2O^+, differing only in the order of the last three vibrational band origins for D_2O^+. The **t** coordinate result for H_2O^+ and D_2O^+ yields sequence (060) < (210) < (111), which is consistent with the internal coordinate result for H_2O^+, but in the case of D_2O^+ the sequence is (210) < (111) < (012). The **t** coordinate assignment appears more consistent with the experimental results. For example, for H_2O^+ and D_2O^+ the seventeenth band origin has an experimental value 7639 and 5650 cm^{-1} respectively compared with the **t** coordinate result of 7631 and 5675 cm^{-1} respectively. The internal coordinate result of Weis et al. [2] gives values of 7662 and 5666 cm^{-1} respectively. Furthermore, for H_2O^+ Weis et al. [2] give the (060) band origin at 7496.6 cm^{-1}. Nevertheless, care must be taken in over interpreting these results for these high lying vibrational band origins, because of basis set incompleteness.

Table 9.8 gives the vibrationally averaged rotational constants spanned over the four low lying vibrational wavefunctions. From least squares fits to the diagonal elements of the H_2O^+ rotational constants spanned by the lowest twenty vibrational wavefunctions the spectroscopic constants ($A_e, A_0, B_e, B_0, C_e, C_0$) using the **t** coordinate solution are (28.82, 28.79, 12.59, 12.41, 8.69, 8.46) cm^{-1} respectively. This compares favourably to experiment values of (27.79, 29.04, 12.60, 12.42, 8.71, 8.47) cm^{-1} respectively, given by Dinelli et al. [11], and with the calculated values of (27.96, 28.77, 12.60, 12.42, 8.68, 8.45) cm^{-1} given by Weis et al. [2].

The limiting case for the rotational levels of all C_{2v} isotopes of H_2O^+ is Mulliken's prolate symmetric top. This is reflected by the Ray's asymmetry parameter which for $H_2{}^{16}O^+$, $D_2{}^{16}O^+$, $T_2{}^{16}O^+$, $H_2{}^{18}O^+$, $D_2{}^{18}O^+$ and $T_2{}^{18}O^+$ has been calculated using the **t** coordinate solution to be -0.612, -0.680, -0.733, -0.602, -0.665 and -0.716 respectively. Hence the rotational levels can be assigned within this framework. Table 9.9 compares the experimental observed rotational levels of the ground and first excited vibrational states of H_2O^+ and D_2O^+ with those calculated using our **t** co-ordinate solution algorithm. The agreement between experiment and theory is excellent, with a maximum deviation of 8 cm^{-1} occuring for J=5, K_a=1 and K_c=5 state of the first excited vibrational level of H_2O^+. Smaller deviations occur for all the other states. For example, on comparison with experiment the calculated rotational levels of the ground and first excited vibrational states of D_2O^+ agree within 2 and 5 cm^{-1} respectively. Similar magnitude of errors are also encountered by the internal coordinate solution by Weis et al. [2]. Hence our calculations confirm the experimental assignment and so give further credence that both the **t** coordinate and internal co-ordinate solution

Table 9.8 Rotational constant and Coriolis Matrix Elements (in cm^{-1}) of the C_{2v} Isotopes of H_2O^{+} [a].

i j	$H_2{}^{16}O^+$	$D_2{}^{16}O^+$	$T_2{}^{16}O^+$	$H_2{}^{18}O^+$	$D_2{}^{18}O^+$	$T_2{}^{18}O^+$
			(a) A'			
1 1	28.7944	15.8875	11.6751	28.4336	15.6206	11.3199
2 1	7.2378	3.2951	2.1800	-7.1343	3.2377	-2.1081
2 2	32.7619	17.3865	12.5701	32.3386	17.0936	12.1841
3 1	1.8454	0.7825	0.5121	-1.8012	0.7545	0.4819
3 2	12.1008	5.2580	3.4113	11.9195	5.1644	-3.2961
3 3	38.0600	19.2087	13.6189	37.5460	18.8833	13.1954
4 1	-2.8057	-1.4490	-1.0510	2.7412	1.3985	0.9934
4 2	0.5180	0.2274	0.1609	0.4848	-0.2062	0.1432
4 3	0.5827	0.2170	0.1358	0.5487	-0.2003	-0.1233
4 4	28.0332	15.6609	11.5773	27.6660	15.3831	11.2131
			(b) B'			
1 1	12.4109	6.2368	4.1938	2.4118	6.2808	4.1942
2 1	-1.2899	-0.5657	-0.3550	1.2845	-0.5662	0.3516
2 2	12.4455	6.2573	4.2083	12.4450	6.3004	4.2080
3 1	0.3807	0.1497	0.0895	-0.3756	0.1478	0.0872
3 2	-1.8146	-0.7983	-0.5008	-1.8073	-0.7992	0.4963
3 3	12.4383	6.2688	4.2187	12.4372	6.3113	4.2177
4 1	-1.2083	-0.4927	-0.2889	1.2148	0.5023	0.2932
4 2	-0.0295	-0.0397	-0.0379	-0.0199	0.0328	-0.0324
4 3	0.0037	0.0097	0.0092	0.0014	-0.0080	-0.0078
4 4	12.1563	6.1374	4.1339	12.1607	6.1829	4.1362
			(c) C'			
1 1	8.4622	4.4016	3.0422	8.4303	4.4021	3.0173
2 1	-0.0089	-0.0408	-0.0454	-0.0005	-0.0330	0.0388
2 2	8.5125	4.4157	3.0489	8.4815	4.4169	3.0242
3 1	0.1321	0.0598	0.0401	-0.1279	0.0572	0.0376
3 2	0.0190	-0.0487	-0.0595	0.0321	-0.0376	0.0502
3 3	8.5797	4.4328	3.0565	8.5500	4.4349	3.0322
4 1	-0.8381	-0.3691	-0.2305	0.8347	0.3700	0.2292
4 2	0.0092	-0.0043	-0.0076	0.0108	0.0025	-0.0057
4 3	0.0169	0.0067	0.0045	0.0161	-0.0062	-0.0040
4 4	8.2844	4.3356	3.0039	8.2533	4.3363	2.9797

Table 9.8 (Cont.)

(d) D'

1 1	0.1573-11	-0.1563-10	-0.1659-10	-0.6368-11	0.4447-11	0.1573-10
2 1	0.3445-12	0.2855-12	-0.5571-12	0.4989-12	0.4470-12	-0.5549-12
2 2	0.1535-11	-0.1629-10	-0.1696-10	-0.5876-11	0.4993-11	0.1593-10
3 1	0.1100-10	0.1360-12	-0.1497-11	0.5141-13	0.7720-13	-0.1100-11
3 2	0.4485-12	-0.1062-11	-0.6005-12	-0.1359-13	0.1327-12	-0.7510-12
3 3	0.2132-11	-0.1845-10	-0.1801-10	-0.6098-11	0.5456-11	0.1739-10
4 1	0.2214-10	-0.6691-10	-0.4251-10	0.5205-10	-0.3694-10	-0.3171-10
4 2	0.1343-01	-0.1574-11	-0.1316-11	-0.1966-11	-0.6820-12	0.1077-11
4 3	0.3552-11	-0.1217-11	-0.6288-12	-0.2158-11	-0.4505-12	-0.5591-12
4 4	0.1092-10	-0.4179-10	-0.2998-10	-0.2848-10	0.1839-10	0.2561-10

(e) F^b

1 1	-0.8163-16	-0.5000-18	-0.1355-16	0.1401-16	0.9887-17	0.3099-16
2 1	0.4319-11	-0.5301-10	-0.6165-10	0.2049-10	0.1535-10	-0.5894-10
2 2	0.1685-15	0.2742-17	0.2500-17	-0.6686-16	-0.2075-16	0.8435-17
3 1	0.7120-11	0.8136-11	-0.1530-12	-0.6218-12	-0.1063-11	-0.9791-12
3 2	-0.9984-10	-0.7592-10	-0.6709-10	-0.2376-10	0.2293-10	-0.9547-10
3 3	0.4929-15	0.6011-17	0.2063-16	-0.1589-15	-0.5346-16	-0.1935-16
4 1	0.1151-10	-0.5387-10	-0.3453-10	0.2859-10	-0.3322-10	-0.2506-10
4 2	-0.2164-09	0.7402-09	0.5076-09	0.4978-09	0.4129-09	-0.3856-09
4 3	0.1842-10	-0.7437-10	-0.3819-10	-0.5401-10	-0.4561-10	-0.3084-10
4 4	-0.9081-16	0.7142-18	-0.7956-17	0.1440-16	0.3503-17	0.3076-16

a) Reproduced with permission from reference [15].

b) Note: $F(i,j) = -F(j,i)$.

Table 9.9 . Comparison of Experimental with Calculated Low-Lying Rovibrational States of $H_2{}^{16}O^+$ and $D_2{}^{16}O^+$ (in cm^{-1})[a].

(a) $H_2{}^{16}O^+$

$J K_a K_c$	This Work	Expt.[b]	Δ	This Work	Expt.[b]	Δ
		$(000)^c$			$(010)^c$	
1 0 1	20.88	20.86	0.02	21.13	20.76	0.37
1 1 1	37.22	37.19	0.03	41.37	41.20	0.17
1 1 0	41.17	41.10	0.07	45.14	45.30	-0.16
2 0 2	61.99	61.96	0.03	62.89	61.73	1.16
2 1 2	75.07	75.09	-0.02	79.88	78.78	1.10
2 1 1	86.87	86.85	0.02	91.16	91.10	0.06
2 2 1	135.44	135.67	-0.23	151.01	151.97	-0.96
2 2 0	136.10	136.29	-0.19	151.51	152.52	-1.01
3 0 3	122.13	122.12	0.01	124.29	121.81	2.48
3 1 3	131.40	131.48	-0.08	137.32	134.63	2.69
3 1 2	154.94	154.89	0.05	159.86	159.23	0.63
3 2 2	198.24	198.47	-0.23	214.59	214.63	-0.04
3 2 1	201.21	201.48	-0.27	216.87	217.32	-0.45
3 3 1	287.76	288.76	-1.00	321.09	324.01	-2.92
3 3 0	287.91	288.82	-0.90	321.18	324.05	-2.87
4 0 4	199.97	200.03	-0.06	204.24	199.68	4.54
4 1 4	205.88	205.98	-0.10	213.35	208.42	4.93
4 1 3	244.64	244.59	0.05	250.49	249.12	1.37
4 2 3	281.24	281.53	-0.29	298.76	297.49	1.27
4 2 2	289.81	289.90	-0.09	305.55	305.09	0.46
4 3 2	372.87	373.66	-0.79	407.12	408.65	-1.53
4 3 1	373.11	374.09	-0.98	407.18	408.98	-1.80
4 4 1	493.49	495.62	-2.13	549.38	554.39	-5.01
4 4 0	493.67	495.63	-1.96	549.52	554.39	-4.87
5 0 5	294.64	294.76	-0.12	301.70	294.27	7.43
5 1 5	298.06	298.22	-0.16	307.64	299.74	7.90
5 1 4	354.95	354.94	0.01	362.53	359.88	2.65
5 2 4	384.18	384.37	-0.19	403.25	399.93	3.32
5 2 3	401.86	401.89	-0.03	417.51	416.37	1.14
5 3 3	478.95	479.66	-0.71	514.33	514.29	0.04
5 3 2	480.85	481.31	-0.46	515.63	515.56	0.07
5 4 2	600.05	601.90	-1.85	657.22	660.42	-3.20
5 4 1	600.42	601.94	-1.52	657.63	660.45	-2.82
5 5 1	750.27	753.37	-3.11	831.80	837.02	-5.22
5 5 0	750.61	753.38	-2.77	832.08	837.02	-4.94

Table 9.9 (Cont.)

(b) $D_2^{16}O^+$

J K_a K_c	This Work	Expt.[d]	Δ	This Work	Expt.[d]	Δ
		$(000)^c$			$(010)^c$	
1 0 1	10.64	10.62	0.02	10.74	10.60	0.14
1 1 1	20.28	20.28	0.00	21.85	21.83	0.02
1 1 0	22.12	22.10	0.02	23.63	23.73	-0.10
2 0 2	31.68	31.65	0.03	32.00	31.59	0.41
2 1 2	39.74	39.78	-0.04	41.55	41.21	0.34
2 1 1	45.23	45.23	0.00	46.88	46.90	-0.02
2 2 1	74.02	74.25	-0.23	80.00	80.59	-0.59
2 2 0	74.26	74.48	-0.22	80.20	80.81	-0.61
3 0 3	62.67	62.65	0.02	63.41	62.53	0.88
3 1 3	68.77	68.84	-0.07	70.96	70.08	0.88
3 1 2	79.73	79.72	0.01	81.62	81.44	0.16
3 2 2	105.98	106.27	-0.29	112.26	112.55	-0.29
3 2 1	107.12	107.39	-0.27	113.21	113.63	-0.42
3 3 1	158.20	159.05	-0.85	171.27	173.00	-1.73
3 3 0	158.24	159.06	-0.82	171.30	173.02	-1.72
4 0 4	103.02	103.04	-0.02	104.46	102.88	1.58
4 1 4	107.24	107.33	-0.09	109.96	108.30	1.66
4 1 3	125.37	125.35	0.02	127.60	127.16	0.44
4 2 3	148.36	148.68	-0.32	155.05	154.99	0.06
4 2 2	151.69	151.91	-0.22	157.89	157.98	-0.10
4 3 2	201.33	202.16	-0.83	214.75	216.04	-1.29
4 3 1	201.41	202.29	-0.88	214.79	216.15	-1.36
4 4 1	272.89	274.55	-1.66	295.37	298.64	-3.27
4 4 0	272.94	274.55	-1.61	295.41	298.64	-3.23
5 0 5	152.30	152.37	-0.07	154.68	152.11	2.57
5 1 5	154.96	155.09	-0.13	158.40	155.70	2.70
5 1 4	181.78	181.79	-0.01	184.58	183.72	0.88
5 2 4	201.04	201.36	-0.32	208.28	207.41	0.87
5 2 3	208.15	208.32	-0.17	214.39	214.20	0.19
5 3 3	255.21	256.05	-0.84	269.05	269.80	-0.75
5 3 2	255.79	256.55	-0.76	269.50	270.24	-0.74
5 4 2	326.85	328.48	-1.63	349.80	352.47	-2.67
5 4 1	326.94	328.49	-1.55	349.90	352.48	-2.58
5 5 1	417.35	419.24	-1.89	451.12	455.78	-4.66
5 5 0	417.43	419.24	-1.81	451.19	455.78	-4.61

a) Reproduced with permission from reference [15].

b) See Lew [9].

c) Vibrational state.

d) See Lew and Groleau [12].

algorithms have both converged for the rotational levels of the ground and first excited vibrational states of H_2O^+ and D_2O^+.

§.35. C_S Case: $KLiNa^+$.

There is little ab initio data on ground electronic state of $KLiNa^+$. In fact, the first exploratory discrete electronic surface was determined by Searles and von Nagy-Felsobuki [16] using the SDCI/FC methodology. The single reference determinant is generally valid for small displacement from the PE minimum, since the leading coefficients tend to be large and constant in this region. For example, at the SDCI level of the theory accurate theoretical spectroscopic parameters have been calculated for the positive ions of the alkaline-earth oxides, fluorides and hydroxides [24]. Invoking the frozen-core approximation is less satisfactory since it circumvents the core and core-valence polarisation effects within the CI scheme and so limits accuracy of the potential function.

All ab initio electronic calculations were performed using the GAUSSIAN 88 [21] suite of programmes. The lithium basis set used was the [11s3p1d/6s3p1d] basis of Gerber and Schumacher [6] with the partially optimized d exponent of 0.15. For sodium, the Huzinaga et al.[22] [16s9p] primitive basis was supplemented with diffuse d and f polarization functions (both with a partially optimized exponent of 0.16) yielding a [16s9p1d1f/10s6p1d1f] contracted basis. For potassium the Huzinaga and Klobukowski [25] [20s13p] primitive basis was used, which was supplemented with d and f polarization functions (with partially optimized exponents of 0.16 and 0.09 respectively) giving a [20s13p1d1f/15s8p1d1f] contracted basis.

For $KLiNa^+$ a 77 point discrete PE surface was constructed [16]. The molecule possess C_S symmetry which yields a minimum energy of -768.36472 E_h and equilibrium bond lengths R_{LiNa}, R_{LiK} and R_{NaK} of 3.073, 3.870 and 4.279 Å respectively. The minimum energy linear form occurs with the lithium atom between the potassium and sodium atoms, in the form K - Li - Na^+, with an energy of -768.35643 E_h. In this case R_{LiNa} and R_{LiK} bond lengths are 3.171 and 3.956 Å respectively. The SDCI/FC calculations predict (symmetry permitting) that the apex angle becomes more acute on substitution of a more electron-dense atom. That is, the sequence for the acute angle is : $Li_2Na^+ < LiNa_2^+ < KLiNa^+$ which is in agreement with a pseudopotential-CI calculation [7].

The potential energy surface of a triatomic molecule can be expanded in terms of internal coordinates. Kuchitsu and Morino [26] have developed a potential expansion in terms of two instantaneous bond lengths and the included bond angle for a triatomic molecule of C_S symmetry.

Table 9.10 gives the force constants derived from the internal coordinate expansions of their discrete ab initio PE surfaces [17]. Such anharmonic force fields have been traditionally utilised in order to examine the physical significance of the higher-order potential constants. In rovibrational calculations it is vital to obtain force fields that accurately interpolate the surface between calculated points. The $(\chi^2)^{1/2}$ using the internal coordinate force fields is 0.1304 E_h and therefore is not accurate enough as interpolating functions in rovibrational calculations. Hence in order to obtain accurate analytical representations of the discrete PE surfaces, various power series expansions and Padé approximates were examined. Six expansion variables were used to form both power series expansions and Padé approximates. The expansion variables used were the Dunham, SPF, Ogilvie and their Morse variants [16]. The "best" representation of the SDCI/FC surface is given by a P(3,3) rational function using an Ogilvie expansion parameter. The coefficients for this fit energy contour plots of the surface are given in reference [16]. The $(\chi^2)^{1/2}$ of the fit is 4.4×10^{-5} E_h. From Table 9.10 it is obvious that the Padé force field is 4 orders of magnitude more precise than the more physically appealing internal coordinate force fields. Thus the Padé force field was used in the subsequent rovibrational calculations.

Embedding this potential function in the f coordinate vibration Hamiltonian the vibrational eigenenergies and band origins of $KLiNa^+$ were calculated and are shown in Table 9.11. The zero-point energy of the molecule is calculated as 173.6 cm^{-1}. Due to the low symmetry even the first three excited vibrations are heavily mixed, with the calculated dominant configuration being due to excitation of the f_1 (breathe), f_2 (bend) and f_3 (asymmetric stretch) modes respectively. The breathe mode overtones are ~56 cm^{-1} apart near the potential energy minimum. For Li_3^+, Li_2Na^+ and $LiNa_2^+$ the breathe mode separations are 299, 165 and 102 cm^{-1} respectively [16], which is consistent with the reduced mass.

The rotational and Coriolis constants given in Table 9.12 are calculated using variational vibration wavefunctions, with the embedded anharmonic force field. It is readily seen that the Coriolis constants are generally very small with the ζ_{32}, ζ_{51}, ζ_{53} and ζ_{54} coupling constants being the exception. The centrifugal distortion constants were calculated using third-order perturbation expansions of the respective operators. Previous work on H_3^+ [3] has clearly demonstrated the convergence at this order.

Table 9.13 gives the variationally calculated vibrational band origins and rovibrational eigenenergies (up to J=5) for the five lowest-lying vibrational states have been calculated up to ~600 cm^{-1} above the zero-point energy. In order to ensure convergence of the calculated eigenenergies with respect to the basis set used, a number of truncated basis sets were investigated. Calculations were employed using five and ten vibrational eigenfunctions respectively. The mean difference of all the rotational levels of the first five vibrational states using these two different vibrational basis sets is 0.1 cm^{-1} respectively. Table 9.13 only gives a small subset of

Table 9.10 Force Constants Derived from an Internal Coordinate Expansion of the Discrete PE Surface of $KLiNa^{+}$ [a].

	Force Const.	Coefficient
	K_4	0.0810
	K_6	-0.0128
	K_7	0.0391
	K_5	0.2971
	K_8	-0.0550
	K_{10}	0.0057
	K_{11}	0.0045
	K_{12}	-0.0136
	K_{13}	-0.0397
	K_9	-0.7649
	K_{14}	0.0206
	K_{16}	0.0071
	K_{17}	-0.0165
	K_{18}	0.0195
	K_{20}	0.0250
	K_{19}	0.0493
	K_{21}	0.1142
	K_{22}	0.0886
	K_{15}	0.8636
$(\chi^2)^{1/2}/aJ$		0.0299

a) All forces constants are in units of aJ (= 10^{-18} J). The minimum energy is -768.3650 E_h. Reproduced with permission from reference [17].

Table 9.11 Vibrational Band Origins of KLiNa^{+a}.

$v_1v_2v_3$	Sym.	Vibrational Eigenenergies / cm^{-1b}	Vibrational Band Origins / cm^{-1}
000	A$'$	173.64 (92)	
100,001,010	A$''$	230.05 (48, 15, 14)	56.41
010,100	A$'$	283.46 (75, 12)	109.82
200,110	A$'$	286.32 (23, 13)	112.68
110,200,020	A$'$	339.73 (23, 12, 25)	166.09
300	A$'$	343.81 (10)	170.17
001,100	A$''$	356.85 (66, 21)	183.21
020,110	A$'$	391.78 (61, 21)	218.14
120	A$'$	396.10 (16)	222.46
200	A$'$	404.80 (6)	231.16

a) Reproduced with permission from reference [16].

b) The values in parentheses are the weights of the dominant product configurations.

Table 9.12 Rotational Matrix Elements of KLiNa$^+$(/cm^{-1})^a.

Matrix Element		A$'$	B$'$	C$'$	D$'$	F^b
1	1	2.519-1	1.771-1	0.529-1	1.578-1	0.000-9
2	1	-0.385-1	-0.680-1	0.000-9	-1.068-2	5.793-4
2	2	2.346-1	1.445-1	0.530-1	1.529-1	0.000-9
3	1	0.139-1	0.244-1	0.000-9	3.656-3	4.403-3
3	2	0.166-1	0.282-1	-0.001-1	4.641-3	-5.138-2
3	3	2.462-1	1.674-1	0.529-1	1.567-1	0.000-9
4	1	-0.024-1	-0.055-1	0.001-1	-7.529-4	6.435-4
4	2	-0.969-1	-1.714-1	0.001-1	-2.668-2	-7.375-3
4	3	0.208-1	0.375-1	0.000-9	5.823-3	-4.846-3
4	4	2.075-1	0.952-1	0.531-1	1.454-1	0.000-9
5	1	-0.055-1	-0.084-1	0.000-9	-1.508-3	1.645-2
5	2	-0.351-1	-0.617-1	0.001-1	-9.554-3	-1.782-3
5	3	0.484-1	0.849-1	-0.001-1	1.336-2	-1.758-2
5	4	-0.272-1	-0.466-1	0.001-1	-7.578-3	6.937-2
5	5	2.256-1	1.292-1	0.530-1	1.509-1	0.000-9

a) Notation 5.529-1 signifies 0.5529. Reproduced with permission from reference [17].

b) The F matrix elements have the relationship $F(i,j)=-F(j,i)$.

Table 9.13 Vibrational Band Origins[a] and Rovibrational Eigenenergies[b] of Low-Lying Vibrational Excitations (/cm^{-1}) of KLiNa$^+$.

E_v(/cm^{-1})[a]	0.000	56.408	109.821	112.681	166.094
J τ					
1 -1	0.105	0.083	0.098	0.048	0.071
1 0	0.429	0.378	0.412	0.301	0.352
1 1	0.430	0.402	0.421	0.360	0.388
2 -2	0.638	0.515	0.593	0.335	0.445
2 -1	0.640	0.590	0.623	0.514	0.562
2 0	0.963	0.858	0.925	0.706	0.799
2 1	0.963	0.858	0.925	0.706	0.799
2 2	1.610	1.469	1.559	1.268	1.391
3 -3	1.252	1.058	1.173	0.791	0.936
3 -2	1.252	1.058	1.173	0.791	0.936
3 -1	1.338	1.294	1.325	1.222	1.276
3 0	1.923	1.705	1.833	1.405	1.569
3 1	2.244	1.968	2.125	1.592	1.791
3 2	2.244	1.968	2.125	1.592	1.791
3 3	3.205	2.869	3.048	2.427	2.644
4 -4	1.959	1.818	1.910	1.558	1.705
4 -3	1.959	1.818	1.910	1.567	1.705
4 -2	2.601	2.147	2.363	1.567	1.813
4 -1	2.601	2.147	2.363	1.762	1.813
4 0	3.343	2.958	3.107	2.575	2.718
4 1	3.649	3.289	3.533	2.657	2.893
4 2	3.832	3.289	3.533	2.657	2.893
4 3	3.832	3.513	3.638	3.382	3.589
4 4	5.014	4.294	4.417	3.382	3.589
5 -5	2.623	2.423	2.531	2.673	2.104
5 -4	2.623	2.423	2.531	2.673	2.423
5 -3	4.206	3.443	3.667	3.118	2.830
5 -2	4.206	3.443	3.667	3.118	2.830
5 -1	4.624	3.851	4.087	3.326	3.262
5 0	4.624	3.851	4.087	3.326	3.262
5 1	4.638	4.477	4.569	4.071	4.202
5 2	5.828	4.938	4.747	4.145	4.202
5 3	5.828	4.938	5.141	4.145	4.280
5 4	6.735	5.955	5.141	5.211	5.294
5 5	6.735	5.955	5.240	5.211	5.294

a) Vibrational band origins. Reproduced with permission from reference [17].

b) Variational eigenenergies up to J = 5.

calculated results. As shown in Table 9.14 Ray's asymmetry parameter is close to zero (being 0.248) and so K no longer has any definite meaning. Hence the rotational levels are assigned in the usual manner.

Table 9.14 gives spectroscopic constants obtained by least-squares fits to the rovibrational eigenenergies. Generally, high resolution rovibrational spectra are assigned by fitting rovibrational data to reduced Hamiltonians. The definitions of the reduced Hamiltonians have been given Chapter II. Vibrational constants such as the fundamental frequencies, zero-point energies and anharmonic constants could not be obtained (within reasonable precision) using a simple least-squares fitting procedure, due to the "mixed" nature of the configuration wavefunctions [16]. Nevertheless, following the usual prescription [27-28], the rotational constants were obtained from least-squares fits of diagonal elements of the matrices spanned by the same number of vibrational wavefunctions. The quartic centrifugal distortion constants were calculated at the equilibrium structure. The signs and magnitude of these constants in Table 9.14 should be of assistance to experimentalists in the spectroscopic detection of $KLiNa^+$.

Ab initio calculations of the discrete dipole moment surface of the ground electronic state were performed at the Hartree-Fock SCF level using the and GAUSSIAN 88 [21] suite of programmes. The basis set used was that employed in generating the PE surface [16]. A 49 point discrete dipole moment surface was calculated [16] in terms of the rectilinear displacement coordinates. The grid used for the dipole moment surface differed from the electronic energy calculations in order to facilitate a more precise fit to an analytical representation. At each point of the dipole moment surface the geometries were rotated and/or reflected in order to coincide with the coordinate system used in the nuclear Hamiltonian (i.e. in the Eckart framework). Hence the molecule lies in the xy plane, with the origin coinciding with the centre of mass and the positive x axis bisecting the KLiNa angle.

In order to model transition probabilities, band strengths and lifetimes confidence in the accuracy of the dipole moment surface is paramount. Obviously for a triatomic molecule with C_S symmetry there is a permanent dipole moment contribution along both the μ_x and μ_y components at the equilibrium geometry. A regression analysis involving powers of the f coordinates were fitted to the discrete surface. The expansion coefficients given in reference [17] indicate that the surface closely resembles a linear function. Hence, it would be reasonable to assume that the error in the discrete dipole moment surface is uniform over the hypersurface considered and of the order of 0.2 D.

Table 9.14 Spectroscopic Constants of KLiNa^{+a}.

Constants	Coefficients	Constants	Coefficients
Vibrational Constants		**Rotational Constants**	
α_1^A (/cm^{-1})	-0.0385	A_e (/cm^{-1})	0.2429
α_2^A (/cm^{-1})	0.0122	A_0 (/cm^{-1})	0.2519
α_3^A (/cm^{-1})	0.0707	B_e (/cm^{-1})	0.1907
β_{11}^A (/cm^{-1})	0.0041	B_0 (/cm^{-1})	0.1771
β_{22}^A (/cm^{-1})	-0.0033	C_e (/cm^{-1})	0.0472
β_{33}^A (/cm^{-1})	-0.0408	C_0 (/cm^{-1})	0.0529
β_{12}^A (/cm^{-1})	0.0011		
β_{13}^A (/cm^{-1})	0.0122	κ	0.248
β_{23}^A (/cm^{-1})	-0.0193		
α_1^B (/cm^{-1})	-0.0716	A+C (MHz)	8366.29
α_2^B (/cm^{-1})	0.0143	A-C (MHz)	7855.38
α_3^B (/cm^{-1})	0.0509	τ'_{AAAA} (MHz)	-569.30
β_{11}^B (/cm^{-1})	0.0075	τ'_{BBBB} (MHz)	-123.48
β_{22}^B (/cm^{-1})	-0.0065	τ'_{CCCC} (MHz)	543.01
β_{33}^B (/cm^{-1})	-0.0412	τ'_{AACC} (MHz)	6154.92
β_{12}^B (/cm^{-1})	0.0022	τ'_{BBCC} (MHz)	3043.70
β_{13}^B (/cm^{-1})	0.0221	τ'_{AABB} (MHz)	-8805.43
β_{23}^B (/cm^{-1})	-0.0151		
α_1^C (/cm^{-1})	0.0001	**First order centrifugal distortion constants**	
α_2^C (/cm^{-1})	0.0017	D_J (MHz)	615.29
α_3^C (/cm^{-1})	0.0145	D_{JK} (MHz)	-3530.23
β_{11}^C (/cm^{-1})	0.0000	D_K (MHz)	2779.19
β_{22}^C (/cm^{-1})	0.0000	δ_J (MHz)	27.86
β_{33}^C (/cm^{-1})	-0.0063	R_5 (MHz)	3483.01
β_{12}^C (/cm^{-1})	0.0000	R_6 (MHz)	264.35
β_{13}^C (/cm^{-1})	0.0000		
β_{23}^C (/cm^{-1})	-0.0037	**Reduction distortion constants**	
		Δ_J (MHz)	86.60
		Δ_{JK} (MHz)	-358.09
		Δ_K (MHz)	135.74
		δ_J (MHz)	27.86
		δ_K (MHz)	-2397.24

a) Reproduced with permission from reference [17].

The vibrational eigenenergies, wavefunctions and dipole moment functions are utilised in calculating the dipole moment matrix elements, Einstein transition probabilities (A_{nm} and B_{nm}), band strengths (S_{nm}) and radiative lifetimes (τ)[29]. Table 9.15 list these quantities together with the calculated transition frequencies with respect to transitions from the ground vibrational state to nine excited vibrational states. For the C_S point group there are no Raman forbidden transitions unlike Li_3^+ (which has D_{3h} symmetry). Hence the lifetimes of the excited vibrational states of $KLiNa^+$ are small for these transitions compared with Li_3^+ [29].

Using the variational wavefunctions of the rovibrational states and the dipole moment surfaces the dipole transition matrix elements are calculated and so, individual line intensities can be calculated using the usual formulae given in Chapter VIII. Table 9.16 gives the variationally calculated rovibrational absorption line intensities for a selected number of transitions. In the calculation the rotational partition function obtained from the variationally calculated rotational levels was employed. As the intensities were calculated at 300 K the contributions from the vibrational partition function was neglected. Although a portion of the rovibrational transitions should be experimentally accessible using laser spectroscopy no data is currently available. In Table 9.16 the absolute line intensities and the squares of the electric dipole transition matrix elements are given for some of the most intense transitions within the P,Q and R branches between the vibrational ground state and four lowest-lying vibrational states.

Table 9.15 Vibrational Transition Frequencies, Square Dipole Matrix Elements, Einstein Coefficients[a], Band Strengths[a] and Radiative Lifetimes for $KLiNa^{+b}$.

j	i	ν_{ji} /cm^{-1}	μ_{ji}^2 /D^2	A_{ji} /s^{-1}	B_{ji} /10^{16}cm^{3}erg^{-1}s^{-1}	S_{ji} /atm^{-1}cm^{-2}	τ /s
1	0	56.41	0.280	0.0044	14.5578	48.7846	6.26
2	0	109.82	0.300	0.0385	17.4681	113.9667	4.77
3	0	112.68	0.058	0.0015	0.6270	4.1975	5.85
4	0	166.09	0.046	0.0030	0.3925	3.8733	4.36
5	0	170.17	0.007	0.0001	0.0083	0.0842	10.72
6	0	183.21	0.190	0.0704	6.8740	74.8159	8.08
7	0	218.14	0.019	0.0011	0.0651	0.8438	5.68
8	0	222.46	0.004	0.0001	0.0032	0.0420	6.75
9	0	231.15	0.001	0.0000	0.0003	0.0046	11.87

a) Calculated at 300 K.

b) Reproduced with permission from reference [17].

Table 9.16 Variationally Calculated Rovibrational Absorption Line Intensities (at 300 K) for χ States of KLiNa[+a].

v'	J'	τ'	v"	J"	τ"	Branch P,Q,R	v_{AX} [b]	S_{AX} [c]	R_{AX}^2 [d]
0	5	3	0	6	-6	-1	2.31	6.70-5	8.50-1
0	6	0	0	7	-7	-1	2.36	6.97-5	8.49-1
0	6	5	0	7	-5	-1	2.42	8.51-5	1.00+0
0	4	4	0	4	-2	0	2.41	8.19-5	9.45-1
0	5	5	0	5	-5	0	4.11	1.10-4	4.37-1
0	7	7	0	7	-1	0	2.82	9.76-5	8.52-1
0	5	5	0	4	0	1	3.39	8.20-5	4.82-1
0	6	0	0	5	-3	1	2.65	8.87-5	8.58-1
0	7	6	0	6	-6	1	8.10	6.98-4	7.27-1
1	1	-1	0	2	1	-1	55.53	8.50-4	2.08-2
1	3	3	0	4	-1	-1	56.68	1.67-3	3.96-2
1	5	-1	0	6	1	-1	53.41	2.82-3	7.63-2
1	2	-2	0	2	-1	0	56.28	2.13-4	5.08-3
1	4	-4	0	4	1	0	54.58	2.71-3	6.94-2
1	6	-2	0	6	1	0	54.69	2.89-3	7.49-2
1	2	1	0	1	1	1	56.84	3.04-3	7.10-2
1	4	-3	0	3	1	1	55.98	3.04-3	7.37-2
1	7	-3	0	6	3	1	54.05	2.69-3	7.18-2
2	2	-1	0	3	3	-1	107.24	1.42-4	1.06-3
2	4	-4	0	5	2	-1	105.90	1.18-2	9.09-2
2	6	-4	0	7	4	-1	102.79	1.10-2	9.20-2
2	2	1	0	2	0	0	109.78	1.22-2	8.61-2
2	5	5	0	5	2	0	109.23	1.27-2	9.25-2
2	6	-3	0	6	-2	0	108.65	1.23-2	9.09-2
2	3	-1	0	2	2	1	109.54	1.28-2	9.14-2
2	6	-4	0	5	-5	1	111.25	1.20-2	8.35-2
2	7	-7	0	6	3	1	105.78	1.16-2	9.10-2
3	4	-4	0	5	-1	-1	109.61	4.54-4	3.28-3
3	5	-4	0	6	-4	-1	110.22	3.65-4	2.62-3
3	6	1	0	7	0	-1	108.61	4.34-4	3.26-3
3	4	0	0	4	-4	0	113.30	4.23-4	2.84-3
3	5	0	0	5	-1	0	111.38	4.70-4	3.30-3
3	6	5	0	6	-2	0	113.66	4.81-4	3.28-3
3	4	2	0	3	-1	1	114.00	4.70-4	3.12-3
3	5	4	0	4	1	1	114.24	4.78-4	3.20-3
3	7	6	0	6	3	1	113.82	4.58-4	3.15-3

a) Reproduced with permission from reference [17].

b) Difference of eigenenergies between rovibrational states labelled as A and X, where A and X represent the upper and lower rovibrational states. All entries in units of cm^{-1}.

c) Band strengths between rovibrational states labelled as A and X, where A and X represent the upper and lower rovibrational states. All entries in units of $atm^{-1}cm^{-2}$.

d) Square of the dipole matrix element spanned by the ro-vibrational states are labelled as AX, where A and X represent the upper and lower rovibrational states. All entries in units of D^2.

REFERENCES TO CHAPTER IX

1 Meyer W, Botschwina P, Burton PG (1986) J Chem Phys 84:891

2 Weis B, Carter S, Rosmus P, Werner H-J, Knowles PJ (1989) J Chem Phys 91:2818

3 Searles DJ, von Nagy-Felsobuki EI (1991) Vibrational spectra and structure, Durig DR (Ed) Chapt. 4, Vol. 19, Elsevier, Amsterdam

4 Searles DJ, Dunne SJ, von Nagy-Felsobuki EI (1988) Spectrochimica Acta A44: 505

5 Stwalley WC, Koch ME (1980) Opt Eng 19:71

6 Gerber WH, Schumacher E (1982) J Chem Phys 76:3075

7 Pavolini D, Spiegelmann F (1987) J Chem Phys 87:2854

8 Herzberg G (1980) Ann Geology 36:605

9 Lew H (1976) Can J Phys 54:2028

10 Strahan SE, Mueller RP, Saykally RJ (1986) J Chem Phys 85:1252

11 Dinelli BM, Crofton MW, Oka T (1988) J Mol Spectosc 127:1

12 Lew H, Groleau R (1987) Can J Phys 65:739

13 Karlsson L, Mattson L, Jardrny R, Albridge RG, Pinchas S, Bergmark T, Siegbahn K (1975) J Chem Phys 62:4745

14 Wang F, von Nagy-Felsobuki EI (1992) Mol Phys 77:1197

15 Wang F, von Nagy-Felsobuki EI (1992) Aust J Phys 45:651

16 Searles DJ, von Nagy-Felsobuki EI (1992) J Chem Phys 95:1107

17 Wang F, Searles DJ and von Nagy-Felsobuki EI (1992) J Mol Struct 272:73

18 Hermann A, Schumacher E, Woste L (1978) J Chem Phys 68:2327

19 Hermann A, Hoffmann M, Leutwyler S, Schumacher E, Woste L (1979) Chem Phys Lett 62:216

20 Eades RA, Hendewerk ML, Frey R, Dixon DA, Gole JL (1982) J Chem Phys 76:3075

21 Frisch MJ, Head-Grodon M, Schlegel HB, Ragavachari K, Binkley JS, Gonzalez C, DeFrees DJ, Fox DJ, Whiteside RA, Seeger R, Melius CF, Baker J, Martin R, Kahn LR, Stewart JJP, Fluder EM, Topiol S, Pople JA (1988) GAUSSIAN 88 (Gaussian Inc.)

22 Huzinaga S, Klobukowski M, Tatewaki H (1985) Can J Chem 63:1812

23 Carter S, Meyer W (1990) J Chem Phys 93:8902

24 Partridge H, Langhoff SR, Bauschlicher WC (1986) J Chem Phys 84:4489

25 Huzinaga S, Klobukowski M (1988) J Mol Struct (Theochem.) 167:1

26 Kuchitsu K, Morino Y (1965) Bull Chem Soc Japan 38:814

27 Wolfsberg M, Massa AA , Pyper JW (1970) J Chem Phys 53:3138

28 Kroto HW (1975) Molecular rotation spectra, John Wiley & Sons, New York

29 Searles DJ, Dunne SJ, von Nagy-Felsobuki EI (1988) Spectrochimica Acta A44: 985

Springer-Verlag
and the Environment

We at Springer-Verlag firmly believe that an international science publisher has a special obligation to the environment, and our corporate policies consistently reflect this conviction.

We also expect our business partners – paper mills, printers, packaging manufacturers, etc. – to commit themselves to using environmentally friendly materials and production processes.

The paper in this book is made from low- or no-chlorine pulp and is acid free, in conformance with international standards for paper permanency.

Editorial Policy

This series aims to report new developments in chemical research and teaching - quickly, informally and at a high level. The type of material considered for publication includes:

1. Preliminary drafts of original papers and monographs

2. Lectures on a new field, or presenting a new angle on a classical field

3. Seminar work-outs

4. Reports of meetings, provided they are
   a) of exceptional interest and
   b) devoted to a single topic.

Texts which are out of print but still in demand may also be considered if they fall within these categories.

The timeliness of a manuscript is more important than its form, which may be unfinished or tentative. Thus, in some instances, proofs may be merely outlined and results presented which have been or will later be published elsewhere. If possible, a subject index should be included. Publication of Lecture Notes is intended as a service to the international chemical community, in that a commercial publisher, Springer-Verlag, can offer a wider distribution to documents which would otherwise have a restricted readership. Once published and copyrighted, they can be documented in the scientific literature.

Manuscripts

Manuscripts should comprise not less than 100 and preferably not more than 500 pages. They are reproduced by a photographic process and therefore must be typed with extreme care. Symbols not on the typewriter should be inserted by hand in indelible black ink. Corrections to the typescript should be made by pasting the amended text over the old one, or by obliterating errors with white correcting fluid. Authors receive 50 free copies and are free to use the material in other publications. The typescript is reduced slightly in size during reproduction; best results will not be obtained unless the text on any one page is kept within the overall limit of 18 x 26.5 cm (7 x $10^1/_2$ inches). The publishers will be pleased to supply on request special stationary with the typing area outlined.

Manuscripts should be sent to one of the editors or directly to Springer-Verlag, Heidelberg.

Lecture Notes in Chemistry

For information about Vols. 1–22
please contact your bookseller or Springer-Verlag